by the same author

DECISIONS, DECISIONS
(with John Coplans)

HAVING IT YOUR WAY

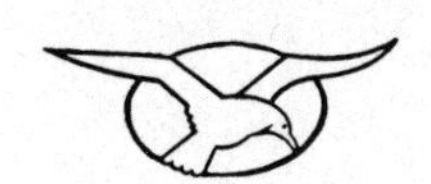

HAVING IT YOUR WAY

The Strategy of Settling Everyday Conflicts

Robert Bell

Drawings by Janet Webb

W · W · NORTON & COMPANY · INC · NEW YORK

Published simultaneously in Canada by George J. McLeod Limited, Toronto. Printed in the United States of America.

Library of Congress Cataloging in Publication Data
Bell, Robert I
Having it your way.
Bibliography: p.
1. Interpersonal relations. 2. Problem solving.
3. Game theory. I. Title.
HM132.B375 1977 158'.2 77-4036
ISBN 0-393-01164-X

This book was designed by Antonina Krass
Typefaces used are Stymie, Florence, and Gael.
Manufacturing was done by Haddon Craftsmen
2 3 4 5 6 7 8 9 0

To the memory of
Hy Bell

Contents

Acknowledgments

Many generous people have offered encouragement or suggestions for this book. I especially want to thank: Jim Alton, Katherine Balfour, Larry Bell, Ricky Bell, Jim Bradley, Ruth Bradley, Michael Brandman, George Brockway, Teresa Brown, Lisl Cade, Arthur Cohen, Jean Colbert, John Coplans, Ralph Collier, John Diebold, Carl Eckels, Barry Farber, Tom Fuchs, Ralph Gardner, Marek Kanter, Elaine Kleinbart, Susan Kleinberg, Owen Lee, Ellen Levine, Alan Lynch, Joel Martin, Ian Masters, Rolf Nelson, Eric Orr, Hank Polonsky, Raquel Ramati, Lynn Russel, Linda Schafer, Willoughby Sharp, Len Stern, Janet Streetporter, Eric Swenson, Marie Torre, Alan Turner, Annina Weber, Ann Wehrer, Mike Wyma.

HAVING IT YOUR WAY

1

Introduction

Despite a scramble for advantage, people in a dispute usually want to settle. But how? And on whose terms? Yours, of course. But can you get your own way and also get a settlement? Game theory, which lays out the strategy of compromise, can tell you.

This book shows how to use game theory to make deals. The focus is on everyday crises in which associations, friendships, families, and marriages hold together or fall apart.

Game theory is usually found on the orderly pages of math books, but a person in crisis would have to apply it in the confusion, emotion, passion, and flux of life. So the use of mathematical formulas seems irrelevant, and their presentation not generous. Why hide behind them?

This book is totally nonmathematical, as a quick flip

through the pages will show. However, the ideas in the book are not stripped down to the point of humbug or oversimplification. The fact is that with regard to bargaining, game theory's message can be explained without using numbers. Some concepts can't be, and they've been scratched. What's left is immediately useful as a guide to action.

* * *

One of the finest private collections of Japanese pottery in the United States, valued at nearly one million dollars, is owned by a man of very modest means. War booty? No, he carefully bought each piece over the years.

"First, I figure out exactly what I'm prepared to spend. Then I tell the dealer, 'This is what it's worth to me.' "

What if he asks for more?

"I walk away."

This book generalizes the idea of a walk-away strategy to include many areas of life where money is not involved. This kind of strategy answers the question, "What's the minimum I'm willing to put up with?" See Chapter 3, entitled "The Walk-Away Strategy."

Such knowledge is important in terms of negative victories. Understanding that a potential deal is worse than walking away can keep one out of emotionally debilitating jobs; the ability to turn down a job may be as valuable as the ability to get one. With regard to marriage, a frequent situation is that of the pregnant unmarried woman who decides to have the baby without either marriage or living with the father. Chapter

6, "Goodbye, Charlie," deals with this issue.

Of course, knowing one's walk-away point implies that one knows what one wants, and many people don't. However, people are often more in touch with their dislikes than their likes. The book exploits this fact to establish priorities, looking first at what one doesn't want, and gradually narrowing the field. Different chapters deal with this idea in various ways, but it is first explained in Chapter 2, "The Practical Idealist."

Have you ever lent money to a friend, and then had trouble getting it back? Of course the Mafia has a standard procedure for this sort of problem. Most people, however, not only want their money back but want to preserve the friendship—and the friend—as well. Just to say that one should speak up is not enough. Exactly what kinds of things should one say? Tact is required, so one must not only figure out one's own priorities, but the other person's as well. Chapter 7, "Don't Take, Have Him Give," focuses on this problem.

The book does deal with speaking up, but with this important reservation: successful agreement involves not demanding too much, and thus getting nothing, and also not getting too much, and so making an enemy. Negotiations are sometimes repeated, and who wants to face someone who has resolved to get even? Avoiding this situation is often the reason for not making an issue of a minor grievance with a colleague. Wounded pride can make a determined enemy, as with the middle-level executive who shows a member of the board to be dead wrong on an issue, and is fired six months later on another pretext.

Speaking up can have other pitfalls, too. The man

who has eyes for his wife's best friend, for example, is in a real pickle. He can do nothing, and be frustrated, or he can do something and run the risk of losing his wife. Let the swine sweat it out? Sure, but what about a more sympathetic version: A young professional wants to change jobs, has spotted a promising firm, but doesn't know how to make the contact without endangering his or her present position. Could using a go-between backfire? What if the executives in the other firm don't want to talk? See Chapter 10, "The Poor Man's Headhunter."

Have you ever been faced with a crisis of conscience on your job? Attorney General Elliot Richardson was, when he publicly resigned rather than fire the Watergate special prosecutor, Archibald Cox. Richardson spoke out in the most dramatic legal way possible. But such public resignations are rare, probably because they are almost always catastrophic for the resigner. Two chapters which deal with this question are "The Practical Idealist," Chapter 2, and "Finessing It," Chapter 4.

They focus on a well-meaning person who works within a system with the intention of changing it. According to game theory, once in a system, the only change is a permutation. Real change means bringing down the underlying standard of behavior, at which point all bets are off as to what will follow.

Although many traditional standards of behavior have broken down, there are now different ones, for example, for women's behavior. But most of the present standards have one thing in common—each person is supposed to place real value on his or her right to speak up, which should not be compromised.

The women's liberation movement involves the change from a discriminatory to a nondiscriminatory standard. In a nondiscriminatory standard of behavior everyone is able to bargain fully, although of course not with equal power, but with an equal right to negotiate.

Evidently, not everyone is committed to the change. A post mortem on the overwhelming defeat of the New York State Equal Rights Amendment showed an amazing fact: far more women than men voted against it. Women want to be discriminated against? Of course not, but some women wanted to be discriminated *for.* Special privileges—such as alimony—are often part of discrimination.

Many women, particularly those who are middle-aged, are caught between the two standards of behavior. On the one hand, a woman may want the dignity that comes with complete liberation. On the other hand, she may well want to continue the known happiness or stability that has been hers throughout a successful marriage but in which she has been the junior partner.

A common situation is that of a wife who, after years of caring for the children, wants to return to college to take up her original dreams for a career. However, her husband may object strongly. What should she do? Chapter 5, "Giving Ms. van der Rohe the Business," focuses on this problem.

Chapter 5 also discusses the nature of compromise. Who should make the concession? How much should be given up? A compromise is simply a certain kind of agreement. What makes any agreement worthwhile?

For a business deal, money, of course. If the money in the deal is more than what both people can get without the deal, it's obviously good. If not greater, the deal's worthless. The people may, of course, make a mistake and go ahead with it anyway, but they'll soon learn about their error.

Suppose the deal isn't about money. Compare each person's walk-away value with the satisfaction from the potential agreement. A good deal is anything better than the walk-away value—whether the issue concerns money, love, marriage, you name it. This is the principle which will be used throughout most of the book.

"We never do what I want to do."

Are there a lot of fights in your family? Are they ever about money? Business has always been about chasing a buck, but family life has only recently been about spending it. The family used to be concerned with producing—the family farm or store where the members worked together to subsist, and, if possible, make money. Now, because of a changed economy, and increased standards of living, the family is most concerned with consuming.

A fixed amount has to be divided and the fewer it goes to the more each gets, as in Westerns, where the characters cooperate until they find the gold, and then shoot it out over the split. Therefore, at least in terms of its economic situation, the typical family is inherently competitive. A classic case is the argument over where to go on vacation, with Dad wanting to go here, Mom there, and the kids somewhere else. Game theory shows that in situations of this kind, there is no single outcome which is better for everyone than all the others. No matter how equitable the arrangement, there is always the potential to disturb it by offering someone a little more of his or her own way and at someone else's expense. Contrary to advertising, the family that goes to movies together, goes to restaurants together, goes on vacations together, may also be the family in which arguments are incessant.

The better bargainers often get their own way. But the less effective ones don't do as well. The aged are in this bind. They did better in the (not so) good old days, when family life was not about speaking out, but about following orders. The consequences of the change to a free-wheeling standard of behavior is discussed in Chapter 14, "What Ever Happened to Mother-in-Law Jokes?"

Of course, the emphasis on spending could be compensated by love, which is often contrasted to money as a motivation.

Everybody knows that money talks, but does it tell the truth? When do you know whether or not to believe an offer? When the check clears, of course; but is there an earlier moment? There is, and the principle involved in finding it can be used to test promises that don't necessarily hinge on money—for example, a marriage proposal.

If the same test of truth can be used for a business deal and a marriage proposal, is there a common ingredient connecting business and love? One would normally say, "No." But in fact, even an auction has the common element. The final bidder and the seller both agree that they're better off than they would be if they didn't make the sale. And, the unsuccessful bidders may be unhappy about the outcome but are powerless to do anything about it, much as a broken-hearted lover may be powerless to do anything about the marriage of "his" woman to someone else. The final outcome in either situation is such that no person, group, or subgroup is capable of spoiling it with a better deal. Chapter 11, "I Give You My Word," focuses on the issue.

Sometimes the promise doesn't come from the person who actually owns, or even controls, the goods, but from a middle-man, for example, a real estate broker. People who want to sell their house are sometimes frustrated to the point of distraction by the realtor's inactivity. Is there some way to test his or her promises in advance? Sure, see Chapter 12, "Moving On."

Realtors occasionally use a client's place as the Brand X house to make somebody else's home look good. This is, of course, a double-cross, which is a potential problem in any deal. Another example is an agreement between actors to audition in cameos as a team. One actor may suddenly try to steal the show, with the intention of looking good at the expense of the partner, whose timing is suddenly thrown off. Usually, directors don't like these antics, and neither of the two gets hired.

A double-cross is only one kind of maneuvering strategem. Another can be simple weakness, which may appear in the short run to act as a strength. For example, one may be able to send messages but not receive them, as with the husband who cannot be contacted because he is closeted in a business meeting. He has his secretary phone his wife to ask her to meet him at the hockey game. She can't stand hockey, but wants to see her husband—perhaps to inform him of their divorce because of his high-handed ways. So even if the baseball manager, Leo Durocher, is right when he says that "nice guys finish last," bastards may not always finish first. Maneuvering often backfires.

The successful maneuver locks its victim into the position the schemer wants. For example, accountants sometimes hold onto their clients for years because they know where the bodies are buried. Of course, not everyone will knuckle under to a threat, just as not everyone has his or her price. Many persons have died demonstrating this point, often to the surprise of those who made offers which could not be refused, but were.

Threats usually involve a comparison, as in the threat of the parent, "It's going to hurt me more than it will you," which means, "I'm perfectly happy to suffer." The threat to litigate in a divorce settlement implies not only an ugly courtroom battle, but an erosion in the value of the community property due to legal costs. Such an erosion may indeed "hurt me more than it will you." The subject is taken up in Chapter 8, "Let's Be Civilized."

So far in this book all agreements have been backed up by an implicit or explicit walk-away threat. But this chapter shows that compromise is possible even in the absence of mutual affection or concern. The fear of damage may be as compelling as the fear of isolation. This is the kind of agreement shown in Westerns—the poker players keep each other honest by exposing loaded forty-fives. All agreements are backed up by one or the other kind of threat.

One kind of walk-away threat comes from someone who suddenly gets cold feet. For example, two people plan a business venture. At the moment the long-sought bank loan becomes available, one partner says, "I'm not sure if I really want to go through with this; I'm of two minds." The other, counting on the project, has to decide what to do. Should he agree to the delay? Should he volunteer some last-minute concessions in an effort to tip the scales? Chapter 9, "Make up His Mind," shows one decisive way to answer the threat.

Office politics are simply a form of two against one—coalitions. Chapter 13, "Who's on Your Side?", shows the hidden source of any coalition's power, and why the sudden appearance of a coalition can be so confusing.

What do you do if you're the one left out in a two against one game? Read Chapter 15, entitled "Left Out," which shows how to keep up the fight.

The original book on this subject is *The Theory of Games and Economic Behavior,* by John von Neumann and Oskar Morgenstern. The authors, both refugees from Europe, showed that the same mathematics which could be used to describe such parlor games as chess or poker could also be used to describe economic and social situations. Two-thirds of their book was devoted to analyzing cooperative games.

However, in the thirty-four years since its publication, there has been no book specifically applying this area of the theory to everyday cooperative situations. The book you are now reading, hopefully, will help to fill this gap. To do so, this book reaches beyond the original theory to include more recent developments.

2

The Practical Idealist

Bargaining usually requires compromise. If an arrangement favors one person, then the other has compromised. If fair to both, then they both have probably made some concession. But sometimes the first compromise to be reached is with oneself—reconciling conflicting motives. This chapter shows how game theory is helpful in settling internal conflicts.

* * *

The president's cabinet has a fast turnover. Heads of departments commonly quit after a year or two, often because they disapprove of a presidential policy. Rarely, however, do the secretaries go public, announcing their objections to the press.

Even if subsequent events show the resigner to have been right, he still often holds his tongue, sometimes for years. By doing so, he remains within the "club," and doesn't "talk out of school."

Why the silence? One reason is worry about future job prospects. Most cabinet and top-level subcabinet officials are recruited from major Wall Street law firms or from the presidencies of the largest corporations and institutions. If they go public on an issue, they fear, probably correctly, that colleagues back in their own industries won't trust them anymore: "If he betrayed the president, what'll he do to us?"

Money may not be the worry. Many resigners have substantial stock holdings and investments. But the exercise of power may be the person's whole life, and at stake is the threat of a permanent loss of power—never again a big shot, a captain of industry. Even on the gravest issues of war and peace the resigner still doesn't go public, but quits instead "on doctor's orders." Reading the papers, one gets the impression of living in a medical dictatorship, the good doctors ordering resignations right and left. And they can't change this impression; doctors are ethically and legally required to keep their mouths shut unless the patient okays an announcement.

Not that the resignations have come easily. Many officials have wrestled with their consciences for months before resigning. Commentators rightly cry "shame," over the resulting silence, which is, in effect, a conspiracy against the public good. And one may fantasize, "If I were there, I'd shout to high heaven about what's going on."

However, quiet resignation occurs not only at the loftiest heights of government, but at humbler places as well. An example is Fred, a painter whose work is shown by an important New York gallery, and who, for the past three years, has been chairman of the art department at a new campus of the state university. Fred cannot live off the sale of his work, and thus needs a job to make up the difference. Teaching would seem to be ideal, giving him the free time he needs to paint. However, his job has not worked out.

Fred objects to the policies of the dean of fine arts and of the majority on the personnel and budget committee. They have been allocating money to increase the physical facilities of the art department, while Fred wants to increase the human resources. He would like to bring in prominent artists, critics, and museum curators to teach for five or six weeks at a time, enticing them with attractive salaries. He feels that this would be of immense benefit to the students, most of whom hope to establish themselves in the art world, and who could, under his plan, make invaluable contacts.

The visiting big shots would inevitably upstage the regular faculty, who, naturally, want to promote themselves. Most of them, however, can't get prestigious commercial galleries to show their work. For this reason they have established at university expense an elaborate university gallery, which has a full-time director and publishes extravagant color catalogues of exhibitions of the faculty's work. In addition, expensive equipment has been bought for making sculpture, ostensibly for student use, but, in fact, largely monopolized by the faculty. In one case, some very costly technical equipment was acquired on the whim of a senior faculty member, and has been sitting unused in storage ever since. Thus, money which could have been used, in Fred's eyes, to enrich the curriculum has instead been spent to advertise and aid the faculty and to enhance the dean's public image.

Because of the disagreement, the locked-door faculty meetings have become shouting matches. However, there isn't the slightest *public* hint of a split. The students know that some of the faculty do not like the chairman, and vice versa, but this has been attributed to "ego trips."

The students should know better because Fred once tried to enlist their support for his cause. He brought in, at his own expense, an outside artist. The idea was to get the students interested in more such visits; *they* would then pressure the dean for the program. Fred even suggested to them that they do so. But they didn't, and Fred couldn't afford to pay for any more visitors.

Although Fred has been battling what he sees as corrupt expenses throughout his tenure in office, he has, as chairman, signed every proposed budget. This hypocrisy so upsets him that he considers resigning his chairmanship, which has another two years to run.

He is tempted to go public, blowing the whistle on the dean and the art faculty in the student newspaper and the faculty senate. But he hesitates; he realizes how difficult, if not impossible, his subsequent position at the university might be. Although he has tenure, and thus cannot be fired, he might not want to stay and take the heat. To look for another teaching job would inevitably bring in the dean, who would certainly be consulted by the new dean at the new college. Fred fears he might be branded nationwide as a trouble-maker and tattletale.

Fred will use game theory to help him handle his problem, including the decision as to what he should say to the dean, and how he should say it.

When applying game theory, the first step is to list as many options as one can think of. Questionable strategies should be listed anyway—they can always be thrown out later. The important point is not to reject anything out of hand.

Fred decides that he has four options: first, resign the chairmanship quietly, perhaps looking for another job, perhaps not; second, go public, blowing the whistle to the student newspaper and the faculty senate; third, suffer through his present job; fourth, resign ambiguously, telling the dean he's resigning for reasons of health, but privately telling others the real reason. However, he would not take the issue to the faculty senate or the student newspaper, but would, instead, leave it to someone else to pick up the ball.

In most situations, including Fred's, the outcome depends not only on what we do, but on the actions of others as well. Our lives are interdependent, whether we like it or not. Therefore, we must consider the options of other people.

Against each of his four strategies, Fred figures that the dean has two basic choices. He may seek revenge in some way, or do nothing, including not badmouthing Fred if he tries to change jobs.

The combinations of Fred's and the dean's options give all possible *outcomes* of the situation. Altogether there are eight for Fred to consider, listed here in no particular order, see Figure 1.

Once you know your options, those of the other party, and, therefore, all of the possible outcomes, you

Figure 1
Fred blows whistle
Fred slips out of
chairmanship quietly
Fred quits but
weasel-words it
Fred sticks out
his job
dean does nothing
Fred blows whistle
Fred slips out of
chairmanship quietly
Fred quits but
weasel-words it
Fred sticks out
his job
dean gets even

need to establish your priorities. This is done by ranking the outcomes according to your preferences. However, people are sometimes more in touch with their dislikes than their likes. So you may find the process easier if you begin with what you do *not* want, gradually working to the preferred outcomes.

Fred establishes his priorities in a discussion with his wife, who asks: "What about staying on as chairman?"

"The last thing I want to do. The frustration and hassle's not only driving me nuts, but keeping me from painting."

"I know."

"The students need to be around some real artists, and it wouldn't hurt the faculty to see some either. But, even though I'm chairman, there's nothing I can do to change the policy."

"And the dean will continue to be a problem."

"Screw the dean, I can't take it anymore."

"Then quit. But haven't you done enough already? Do you have to bring up the case with the faculty senate? Tell people privately how you feel, and let someone else take up the fight for a change."

"But I'd feel like a sneak."

"You think it's copping out?"

"Obviously it is, and I don't care what the dean might do, it would simply be an awful way for me to behave."

"But it's better than staying on as chairman."

"One notch better, but still bad. My point is: I can't stand people who do that sort of thing, and I don't want to be one of them."

"But if you blow the whistle, you're just asking for trouble."

"That's for sure. But you know, I wouldn't feel so bad about it even if I had to take the heat. Who knows? Maybe I could change the policy. Maybe the *dean* would quit."

"Well, you could keep your job as faculty member regardless of what happens."

"Even the dean can't do anything about that—I'm tenured. The problem comes up if I want to leave the Big U and teach somewhere else. Suppose the dean does nothing. Suppose he's whipped. People at other colleges know what's going on. Nobody wants to hire a trouble-maker. The fact that the policy I believe in is right may not make any difference at other places, where the faculty members are probably working their own angles. So even if I'm successful, it's still only so-so."

"Well, what does that leave you with?"

"Resigning quietly."

"Yes, that's the best thing."

"Except, I wouldn't have spoken my mind."

"In three years as chairman, you haven't found one person in the art department to back you up. Why should people in other departments give a damn?"

"Well, they might, but you're probably right and this way I'll get back to work. My next painting will be called, *He Who Fights and Runs Away Lives to Fight Another Day.*"

"He does, unless the dean gets nasty for old-time's sake."

"Outsiders would wonder why he's trying to make me miserable. It would be pretty good if he brought the roof down on himself by his own gratuitous vengeance."

"Can you resign quietly and cleanly? As you say, people will ask questions."

"If anybody asks, I'll just say 'Too many headaches!' Frankly, I think the dean'll be glad to get me out of his hair and this way everybody gets off the hook."

"But how can you keep from grumbling?"

"Take next semester off and we go to Europe."

Fred has now established his priorities, see Figure 2.

Fred was able to rank his feelings because he was immediately confronting the issue, not speculating on someone else's problem—he *knew inside* how he felt because he felt it.

Your feelings always accurately reflect you, even if your reasons, your explanations for your feelings, may sometimes be wrong. This is why people can often rank their feelings more easily than they can give an explanation for them, and why they should ignore advice on how they should feel. Such advice is of necessity based on someone else's feelings. Notice also that Fred's morality, not game theory's, is reflected in his priorities. Game theory can be no more moral than the person applying it.

One quality of Fred's priorities should be noted: *transitivity,* or consistency. For example, an item ranked as "awful," although preferred to an item ranked as "the last thing I want to do," is also never preferred to an item ranked as "so-so." In other words, it is strictly inferior to all the items below it on the list. Without transitivity one usually can't make a decision, much less an agreement, because the grass is always greener with another choice. We don't claim that everybody thinks and feels with transitive clarity, but we do say that unless your preferences are transitive, you won't be

Figure 2
THE LAST THING I WANT TO DO.
I stick out the job/regardless of what the dean does
AWFUL
I quit but weasel-word it/regardless of what the dean does
NOT SO BAD
I blow whistle/dean gets even
SO SO
I blow whistle/dean does nothing
PRETTY GOOD.
I slip out of chair-manship quietly/dean gets even
BEST!
I slip out of chair-manship quietly/dean does nothing

able to make either an effective decision or an agreement, since your choice will always be fouled by shifting desires. However, there are cases where even with transitive priorities you are rationally torn between alternatives. This situation, and its consequences for making deals, will be examined in a later chapter.

Another property in Fred's rankings: When he compared any two items the others stayed put. They were irrelevant to his analysis of the two he was considering at that time. If this weren't so, his ranking process would be pure chaos. This is a way of telling himself, "One thing at a time."

* * *

Before analyzing Fred's bargaining strategy, we'll see how the dean feels. He, of course, is well aware of Fred's views, having many times argued with him in private: "Fred, don't you see, these exhibitions put our university on the map—the catalogues are seen by universities and museums all over the world." But Fred has made his feelings clear: "Bullshit!"

The dean establishes his priorities in a discussion with Mike, a full professor in the art department:

"Mike, I want to be fair."

"And we don't want anyone asking why Fred's being fried."

"That would be the worst thing, and if he resigns quietly there's no need for it. The best thing is to let bygones be bygones."

"Why not suggest he quit?"

"Too risky. It might provoke him. My only options are to do nothing, or to make his life miserable, but never to suggest he resign. I have to let him do that entirely on his own."

"The question is, how he does it."

"Well said. If he thinks he's Daniel Ellsberg, I'm going to break him. You can count on that! I'll tell you Mike, getting him would make me feel good, and letting him get away with it would make me feel terrible. That would, indeed, be the worst."

"What can you do if he doesn't quit, but, instead, stays on and continues to be a pain in the neck?"

"It makes no difference what I do. It'll just be bad. We don't have the option of removing him, and his term lasts another two years."

"You know, Oz, he could slip out quietly without a public stink but still make trouble by grumbling to the wrong people."

"And somebody might listen. It could become a problem. Now we haven't done anything illegal, but the chancellor just might decide we have a little too much money to play around with in this division. You know the petty minds of some administrators. They don't always realize that when you're trying to do big things you sometimes make some small mistakes."

"Those are just what I'm worried about. But if you go after him . . ."

"Or not, either way, it's bad for us."

"Yeah."

The dean has established his priorities, see Figure 3.

A comparison between the list in Figure 3 and that for Fred in Figure 2 shows one minor difference. Although Fred's list is in order of priorities, as it happens, his list also is in terms of strategies—each of his strate-

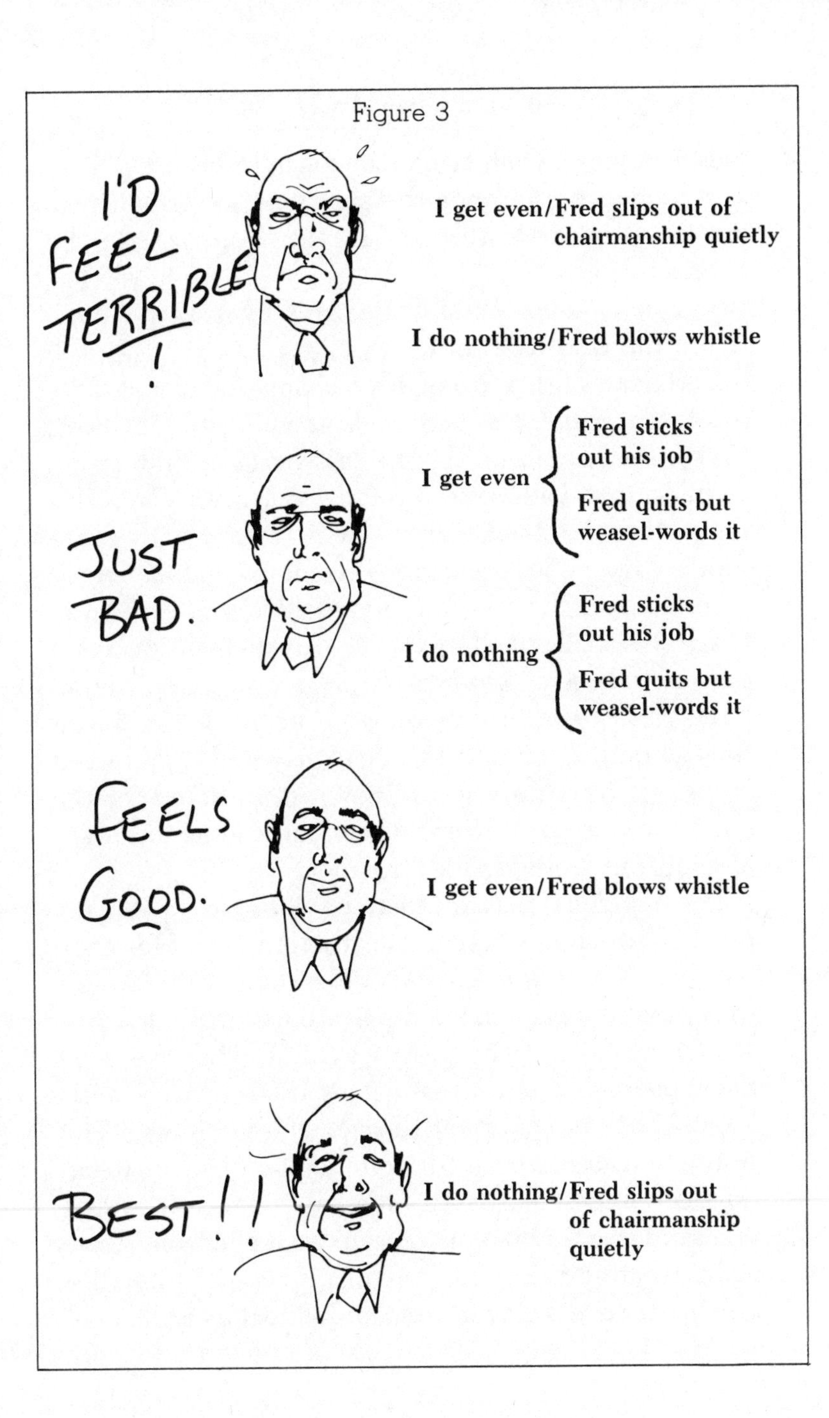
Figure 3
I'D FEEL TERRIBLE!
I get even/Fred slips out of chairmanship quietly
I do nothing/Fred blows whistle
JUST BAD.
I get even
Fred sticks out his job
Fred quits but weasel-words it
I do nothing
Fred sticks out his job
Fred quits but weasel-words it
FEELS GOOD.
I get even/Fred blows whistle
BEST!!
I do nothing/Fred slips out of chairmanship quietly

gies occupies a distinct bloc of space in his list. This is a coincidence, but once the priorities are established, any list can be rewritten in this form. For example, the dean's list, Figure 4.

Why write a list in this way, that is, by strategy? We'll see in the next chapter.

The dean's list of priorities has only four levels to it (I'd feel terrible, just bad, feels good, best). However, Fred's list has six levels (the last thing I want to do, awful, not so bad, so-so, pretty good, best). Does this mean that Fred—the artist—feels things with more subtlety than does the dean? If we looked at his art we might find out, but in terms of priorities, there is no way to tell. We shall not compare the private feelings of one person to those of another. To do so would imply absolutes, and in feelings we know of no absolutes. So, although both Fred and the dean have least-preferred items, we have no way to know if they stand for the same intensity of feelings. If anyone claims there are absolutes in feelings, ask for proof.

The paradox is that in situations where we know people well, we almost never use absolute values for comparisons. The statement, "We love each other," refers to completely personal and private feelings, each person loving in his or her own way. But, when we don't know people well, we treat values as if they were absolute: "That's too good for those welfare chislers." This distinction accounts for the complaint of the wife to her political activist husband: "It's easier for you to love mankind than to love me or your own children." She's right, treating values as absolutes is easier. Otherwise, one has to take a case by case, individual by individual, value by value, approach. Each person would have to be treated on his or her own terms, which is the way this book deals with the matter.

Fred knows his preference ordering, and the dean knows his own as well, but they don't know each other's. That knowledge can only come about through bargaining, interaction between the two which has not yet taken place.

But before they can come to any meeting of minds,

Figure 4

I'D FEEL TERRIBLE!	I get even/Fred slips out of chairmanship quietly
JUST BAD	I get even/Fred sticks out job
	I get even/Fred quits but weasel-words it
FEELS GOOD	I get even/Fred blows whistle
I'D FEEL TERRIBLE!	I do nothing/Fred blows whistle
JUST BAD	I do nothing/Fred quits but weasel-words it
	I do nothing/Fred sticks out job
BEST!	I do nothing/Fred slips out of chairmanship quietly

each must work out his walk-away value, as discussed in the next chapter.

This chapter has shown how to establish one's priorities:

YOUR OPTIONS

First, list the widest range of feasible options, including those you think you would not like to use.

THEIR OPTIONS

Second, realize that we do not live in isolation, and that what happens depends not only on what we do, but on the actions of others as well. Try to list their possible options.

COMBINATIONS

Third, work out the combinations of options. For each of your options, write down every one of the other person's, as in the list in Figure 1. These are the possible outcomes to the situation.

PRIORITIES

Fourth, establish your priorities among the outcomes. Sometimes the easiest way to begin is to pick out the outcomes you least want, and then proceed from there. Complete your priorities as in the lists in Figures 2 and 3.

REARRANGE

Fifth, rewrite your list of priorities, grouping together all outcomes which require the same strategy of yours, as in the dean's list in Figure 4.

PRIVATE FEELINGS

Sixth, and most important, remember that there are no absolutes with regard to feelings. Everyone's feelings are inherently private, and can't logically be compared. If you are making judgments about someone else's feelings, you're always guessing.

3

The Walk-Away Strategy

"What's the absolute minimum you'll accept?"
"This is what it's worth to me."

Both of these remarks refer to a very special price—if not met, at least one person will walk away. This chapter generalizes the concept of a walk-away price to include not only money, but many other things to which value is attached.

* * *

Fred's priorities represent only his feelings, and not the dean's. This single-mindedness is useful. If he and the dean squabble, and Fred walks away in a huff, he can figure out the best he can do by himself, even if the dean is particularly nasty and tries to make Fred as miserable as possible. This would be Fred's *walk-away strategy.*

Notice in Figure 2, p. 34, that Fred's option to "slip out quietly" is at least "pretty good" if the dean gets even, and "best" if the dean does nothing. Thus, if Fred chooses this option, the dean cannot reduce Fred's value below "pretty good." Similarly, with the strategy "blow the whistle" the lesser value is "not so bad." With the strategy "quit but weasel-word it" both values are the same, "awful." With the strategy "stick out the job," both values are again the same, "the last thing I want to do." Since "pretty good" is better than "not so bad," "bad," and "the last thing I want to do", the value "pretty good" is the best Fred can expect if the dean is out for vengeance, and is Fred's walk-away value. Resigning quietly becomes his walk-away strategy, the best he can do if he and the dean end up shouting abuse at each other.

Although now obvious, Fred's walk-away strategy only became so *after* he established his priorities. His life has been a constant upward escalator: college, marriage, graduate school, a teaching job, prestigious art exhibitions, and then the chairmanship. "I never really made any decisions, everything just seemed to happen right," he admitted. The resignation was his first setback, and the fact that he had to make any decision at all was the *real* crisis. Game theory gave him a way to come to grips with it.

Suppose your conscience requires that you "blow the whistle." Could game theory help? Sure, by helping you face the full consequences, it may enable you to know if you're listening to your conscience or to a romantic fantasy. Of course, if you think you're Errol Flynn, that's your business. In any case, although edifying, such an example will not appear in this book, which is about making deals. When going public, one stops negotiating, except perhaps with the D.A.*

What about the dean's walk-away value? Each of his two strategies could result in four different values, depending on what Fred does (see Figure 4, p. 39). The least pleasant value if the dean gets even is "I'd feel terrible," and the least pleasant value if the dean does nothing is also "I'd feel terrible." Either way, the worst that can happen to the dean is to get his absolute worst, which is, therefore, his walk-away value. Unlike Fred, the dean does not have an obvious walk-away strategy —both options look equally bad. The reason we rewrote the list in Figure 3, p. 37, into that in Figure 4, p. 39, may now be obvious. Figure 4 shows, at a glance, the worst outcome for each of the dean's strategies.

Since the dean's walk-away value is so unpleasant, we might suppose he'd be eager to talk things over with his chairman. On the contrary, the dean is scared to open his mouth.

Not knowing how Fred sees things, the dean may fear that Fred is out to make him miserable. (Top administrators are often suspected of such paranoia.) The only way Fred could twist the knife is on the dean's own terms, in other words, based on the dean's priorities. If Fred knew how the dean viewed things—which he doesn't, but ignore this fact for the moment—Fred would choose an option giving the dean the worst possible value.

*For examples of this kind of decision, see *Decisions, Decisions* by Robert Bell and John Coplans (New York: W.W. Norton, 1976).

So the dean would be foolish to tip his hand. If he hints that he plans to "get" his chairman, Fred merely has to resign quietly, reducing the dean's value to "I'd feel terrible." This could happen if the dean badmouths Fred publicly and Fred gets wind of it. Similarly, if the dean lets people know that he will do nothing, Fred merely has to blow the whistle, again reducing the dean's value to "I'd feel terrible." This could happen, for example, if Fred first lines up another job, getting a good reference from the dean, and then blows the whistle. So, the dean may well feel that he should hide out.

In fact, this is exactly what's happening, since Fred and the dean haven't talked to each other privately for months, the dean always finding some excuse to be unavailable, and telling close friends, "I'm keeping my options open."

Deans and other senior administrators are often accused of acting in arbitrary ways. Their objective sometimes seems to be exacting psychological tribute rather than developing good will and smooth administration. A classic example is their busy schedules, which means you see them when they are good and ready. However, their "Wizard of Oz" style may not always be caused by egotistical caprice, but may be due instead to a clear understanding of an often vulnerable situation, in other words, to cold calculation.

So, the dean is in a bind; he would like to talk things over with Fred and get him gently eased out, but he's afraid to start talking. Did the use of game theory put the dean in this kettle of fish? Far from it! The dean's own values put him in the soup; game theory just told him the best way to behave given his values. A different dean, with different values, might not be in the same bind, just as a different chairman, with values different from Fred's, might have blown the whistle for the students' benefit, and have a clear conscience in his new job wrapping packages at Macy's.

* * *

This chapter has shown how to work out your walk-away value, the least you have to settle for if you can't make a deal.

BY STRATEGY

First, make sure that your priorities are listed by strategy as in Figure 4. All outcomes for each of your strategies should be grouped together.

THE DARK SIDE

Second, look at the worst outcome for each of your strategies.

LEAST BAD

Third, your walk-away strategy is the one with the least-bad outcome, which is your walk-away value.

4

Finessing It

Fred's walk-away value is so high that he doesn't really care whether he talks to the dean or not—he could just as easily send a letter informing the dean of his resignation. And the dean's walk-away value is so low, he doesn't want to do any talking! But Fred's life has always been smooth, at least until now, and he would like to make it so again. "I've known guys whose lives were constantly going wrong. Usually it happens when they get off a clear track; I want to get back on that track as fast as possible," he tells his wife. So Fred decides to try squaring things with the dean. After all, he's going to quit without causing a stink, so he may as well do it as pleasantly as possible.

Although perhaps unethical from the students' point of view, Fred's decision will be viewed as highly ethical professionally—not talking out of school being the highest virtue. Fred feels slightly guilty about this when he tells his wife: "Well, if I'm short on scruples, I'd better be long on finesse."

Shortly before the end of the fall semester, Fred takes the initiative: "Oz, you know I've been opposed to the gallery from the beginning. And, I think with another year at it, I could persuade you guys of the wisdom of my position. But the whole job of chairman involves too much tension—it's giving me an ulcer. I have to reluctantly resign, effective the end of the term. Also, I need to get away for a few months, so I'll have to take a leave of absence for next semester. By the way, could you keep my resignation and leave of absence under your hat until term ends?"

"Fred, you've just cured two ulcers, not one. In gratitude for this present of *my* good health, I'll see to it you get next semester off on paid sabbatical, not leave of absence. And, I won't tell anyone except the new interim chairman, Mike."

Everybody's happy. Each gets what he feels is best for himself; they couldn't both have done any better—and the students couldn't have done any worse, but that's *their* fault for not screaming about the corrupt departmental practices. Notice that neither Fred nor the dean ever learned in detail what the other's values were. And neither really cared. Although both are happy, they remain enigmas to each other.

Conclusion: If you have a clear walk-away strategy, which happens also to be your bargaining strategy, and your walk-away value is high, take the initiative and start talking. You've already decided what you're going to do anyway, and by talking you might end up better off. But, if your walk-away value is low, and, in particular, if you do not have a clear course of action, and, thus, tipping your hand is dangerous, keep a low profile, keep your mouth shut, and wait to see what happens. Maybe your problem will go away.

5

Giving Ms. van der Rohe the Business

Love is not all you need in marriage. Money is essential too. So is personal fulfillment, but when a couple is preoccupied with setting themselves up financially, the wife often can't focus on fulfilling her own ambitions. This is like a fixed fight, where one boxer takes a dive in exchange for cash. Both make more money than they would otherwise, but one will never have glory, and the other knows he's cheated.

Sometimes the wife's ambitions are not abandoned completely, but are revived years later, after the children are grown. By this time, however, a pattern of life has been firmly set, and making an adjustment can be wrenching. This chapter is about compromise, and shows one way to make a successful new agreement.

Steve and Lisa, happily married for eighteen years, have two teenage children. They were poor in cash when they married, but had a large capital investment in Steve's education, as he had just completed an advanced engineering degree.

The future turned out to be golden. By making the right decisions at the right time, his small engineering firm grew fast. Lisa played an invaluable part, entertaining clients at dinner parties, and showing her husband to be a reliable family man. Because she loves Steve, and takes pride in him, she has been happy to play this role.

However, she isn't only the hostess with the mostes', but is also included in Steve's business—listening and advising. He never makes any major decision without first thoroughly discussing it with her, and usually following her advice. In this sense, she is a valuable partner, but not, of course, an official one, since she lacks a degree in engineering.

After ten years of marriage, and moderate success in the engineering world, Steve and Lisa set up a joint business, an advertising design agency. Steve is the major stockholder, Lisa the minor one. This business has grown slowly, but significantly. Now it's well known, and has given Lisa money in her own name.

Not that Steve hasn't been generous. On the contrary, he has been lavish, giving Lisa money to do with as she pleases. Lisa rarely needed to spend any of it, since he also bought her anything she wanted. In short, she has what many women would think is the best possible marriage.

When Lisa married, having just graduated college, she willingly laid aside her plans to get a graduate degree in architecture and to become a licensed architect.

This was necessary to help with Steve's career. On Lisa's part, there was no resentment. This happened at a time when many women were giving up promising careers to make room for husbands who were returning from Korea. Although Steve was not actually in the war, he benefited from the spirit of the time.

Over the years, Lisa has nurtured the hope of returning to college to finish her training in architecture. The children no longer need her supervision. In fact, the question is now whether they have time for her, rather than vice versa. Lisa sees her potential career in architecture as a way to reestablish her authority as a mother by being someone whom the children can continue to respect.

She's mentioned the idea to Steve on several occasions. Each time he says, "You might look into it someday."

One day she does. Through Frank, an architect and family friend, Lisa meets the dean of the best local architecture school. She applies and is accepted. Luckily, most of the applicable credits from her original undergraduate degree are accepted, reducing her course to three fulltime years. Lisa is thrilled, but has said nothing to Steve, planning to surprise him with the good news. The next morning she does:

"I've been accepted at architecture school."

"That's nice, too bad you can't go."

"What do you mean I can't go?"

"Do you really want to mess around with that? I know you've had this pipe dream, but here you are—with position, money, two great kids, a good social life, and an interest in the advertising agency—everything you could possibly want. But, now you want to be an architect. What for?"

"You don't want any competition in the family."

"Bullshit."

"Steve, I know how you've talked about architects; they claim to have the ideas, and treat you as the muscle. You're afraid I'm going to do the same."

"Not at all, *Ms.* van der Rohe. I don't want you to go because I need you here when I'm here."

Lisa genuinely feels Steve's pride has been wounded. He has provided her, generously and lovingly, with everything she has ever appeared to want. And now she's saying it's not enough. She also believes she's hit a raw nerve about the architect as the creative brains and the engineer as the muscle. She feels this is one of the reasons Steve got involved in the advertising business—as an outlet where he could be *publicly* recognized as creative. Of course, Steve knows perfectly well that engineering requires a great deal of innovation and imagination, but he is only too aware that many nonengineers don't: "An engineer, not an architect, designed the first steel suspension bridge, and the first railway tunnel under a major river. I am speaking, of course, of Isambard Brunel."

"Who?"

Since Steve's social world is almost entirely outside of engineering circles, his professional achievements are not known to his friends, and among them he feels like a nurse instead of a doctor. Now he feels that even Lisa may be slightly contemptuous.

Having no clear idea of what to do, and needing

someone as a sounding board, Steve talks to his friend, Sid:

"I had no idea she'd actually go through with it."

"Put your foot down. Tell her she can't."

"I sort of tried; it didn't work, probably because I've never opposed anything she's wanted. But the hell of it is, Sid, that I can't simply go along with this and say 'Gee, honey, that's terrific.' Because, I think it's awful. I used to say, 'look into it,' because I never thought she'd actually get around to it."

"Sounds like Lisa hit the nail on the head when she said you don't want an architect for a wife."

"I guess so."

"Steve, ignore it. She'll drop it now that she knows how you feel."

"If she does, great, but she's socially committed. Frank's introduced her to the dean. The whole thing's set. I can't ignore it because it's real and isn't going to go away."

"Make fun of her until she feels too foolish to continue."

"I couldn't do that. If she dropped it, she'd be bitter, and, regardless of what she did, I'd feel like a jerk."

"Buy her off."

"With what? I've already bought her everything she's wanted until now, and what she wants now I can't buy her because it's not for sale."

"What's that?"

"To feel like her own person, not an extension of me."

"Yes. And, she should have it! Give her the business."

"What?"

"The advertising agency. She could become outstanding in her own right, *and,* without crowding you, *and,* with the same social contacts you both have right now."

"Yes! That's an idea. But, what if she takes it and still goes to architecture school?"

"How can she? She can't do two things at once."

"Right."

"Of course, she might go part time."

"Yes, she might."

"Yes."

"Fine. She'd gradually disengage from architecture. To tell you the truth, Sid, I don't think she realizes what she's getting herself into—hanging out with a bunch of college kids. She's used to a much more sophisticated crowd. I mean, what are they going to offer her?"

"The fantasy of being young again. Those kids are on the right side of their disappointments. And at this time, Lisa may feel she's not."

"So you think she might get more attached to school once she starts?"

"Well, school is a way of putting off reality. But advertising is right now. If she really gets into it, it just might become more exciting than doing a lot of mechanical line drawings."

"And, she starts at the top in advertising."

"That's right. But don't sell school short. Present the offer as an idea you had before school came up. Emphasize the immediacy of the satisfaction. If she *is* worried about her age, this might mean something."

Lisa, as confused as Steve, talks things over with Frank:

"Steve is furious; he doesn't want me to go to school. But I just can't drop the idea."

"Why should you? When you mentioned architecture school to him before it was a fantasy, a dream. Now, it's reality. He's probably as surprised as you are about his true feelings."

"I think so."

"Maybe you can get him around with a little time. Could you put it off for a semester while you worked on Steve?"

"No. I'm not getting any younger, why should I wait? Besides, I've been through the whole business of dealing with the dean and the graduate school, and of getting all the records straightened out. I mean, I know the school won't fall apart if I don't go, but if I put it off for a semester, the dean'll think, 'Oh, just another silly daydreamer, incapable of doing anything.' So, I'm not going to put it off. That's out."

"Well, I can understand that. Would you consider going part time?"

"I hadn't even thought of it."

"If you start out part time, Steve might gradually get used to the idea. It wouldn't be as big a threat. Then switch to full time. It's a longer project, but it may be a better idea."

"But I want to get the degree over with."

"He's obviously not going to be happy about it. Do you want him to insist you don't go? What would you do then, wreck your marriage over it?"

"I love Steve, that's the last thing I want."

"Now that you see a possibility of going part time, why don't you talk it over with him again?"

Lisa hasn't talked to Steve enough to know his options, or his feelings, other than that he doesn't favor her plans. In particular, she doesn't know what concessions he might make. So far, all they've done is fight.

Steve and Lisa would both like to settle this thing so that everybody is happy. In order to make sure of this, each must know not only his or her own priorities, but the other person's as well. A good mediator might help to clarify these issues with a minimum of shouting, abuse, and recriminations. But, since they don't have, or want, a mediator, they'll have to spell out their feelings to each other themselves.

This is where the use of priorities is tremendously helpful. By expressing their feelings in terms of priorities, they focus on what they want, rather than on destructive abuse and irrelevant past sore points. Furthermore, if they both want a long-term solution, they must not try to mislead each other for short-term advantage.

That evening at dinner, they talk it over:

"I know it bothers you, but I really intend to go to architecture school."

"What can I say? I can't be a hypocrite about it."

"No, say what's on your mind."

"This morning I tried to put my foot down. It didn't work and I'm glad. I wouldn't want to be like that anyway. The problem is, Lisa, I won't be able to ignore it. It's going to grate on me everytime I'm home and you're at school, or I want to go out and you have to do your homework."

"I know, you've come to expect certain things out of our marriage."

"You were right when you said I resented architects. I can't get over that in ten minutes. What scares me is that you're going to go to school, and I'm going to get bitter. I don't want to be, but I can see it developing."

"That's not very good, not what I want."

"Well, I haven't told you one of the reasons I'm upset: For the past few months I've been very seriously thinking of quitting or significantly reducing my involvement in the advertising business. And, I had been hoping that you would take it over."

"Really?"

"Yes. You know I haven't been to the office in two months."

"I know."

"What I mean is, I had been planning to sign most of my shares over to you. You'll have a clear majority—it'll be your business, and you'll get the profits."

"It's not profits I'm interested in."

"I know, but there's nothing wrong with money."

"It does interest me to be the one who makes the decisions, if that's what you mean."

"That's exactly what I mean."

"Well, that's worth thinking about."

"You've already been making some of the key decisions for the past couple of months."

"But I can't possibly continue to do that and go to school."

"Then forget school . . . or try it out with just a course or two."

"Look, let's lay our cards on the table. The worst for me is either to give up school or put it off; both are out. But, to have you ridiculing me if I go to school part time is also pretty bad. I mean, if you're going to give me static, I may as well go full time."

"Yeah, well if you go part time, I'll try to be reasonable."

"And the best for me is to go full time, without any trouble from you."

"Well, you can't have my share of the business if you go to school full time.

"If I go part time? Exactly what are you offering?"

"I'll sign the shares over to you right now. I think you could handle them both because you're extremely capable."

"How perceptive of you, Steve."

"Yes."

"It might be pretty good to do both."

Lisa's priorities are shown below in Figure 5.

Lisa's walk-away strategy is to go full time, but her walk-away value is "not very good." This is easy to see from a glance at Figure 5. If Lisa "forgets it," all outcomes are "out," and if she delays, it's the same story. She knows from their conversation that Steve won't favor her being a student, but he also won't put his foot down. So she can ignore those issues. With these deletions, the worst that can happen if she goes full time is that Steve will be nasty, which is "not very good." However, this outcome is better than the worst if she goes part time—when Steve again is nasty. This is "pretty bad." She'd rather have an outcome which is "not very good" than one which is "pretty bad." So, going full time is Lisa's walk-away strategy, and her walk-away value is "not very good."

Of course, Lisa could easily size up her walk-away value without making a list. Since she had already decided not to forget about school or delay, she merely

I put it off / I forget it { Steve does anything
THAT'S OUT!
I go full time/Steve puts foot down
THE LAST THING I WANT!
I go full time/Steve is nasty
NOT VERY GOOD
I go full time/Steve offers the business
GOOD!
I go full time/Steve favors school
THE BEST!
WISHFUL THINKING !!!
I go part time/Steve puts foot down
THE LAST THING I WANT!
I go part time/Steve is nasty
PRETTY BAD
I go part time/Steve offers the business
PRETTY GOOD
I go part time/Steve favors school
GOOD!

Figure 5

had to look at her two remaining strategies, and focus on the worst outcome for each. However, she should make up the list anyway, for reasons which will be apparent later.

"Okay, Lisa, you've told me how you feel. Now, it's my turn."

"Okay."

"As I said, if you reduce your involvement, I think I can handle it. I mean, I'd feel like a jerk, otherwise."

"Yes."

"On the other hand, if you turn down the offer and go full time, that's going to grate on me, too."

"Yes."

"It would be as if you're completely cutting yourself off from me and from our life together, and that would be pretty bad. Besides, that ad agency is a good deal."

"I know its merits; I've been working there off and on for years, but not as the boss."

"Yes, and you'd make a good one. Frankly, if you go to school full time, I'd be disappointed in your sense of perspective. I'd try not to be nasty, but I might be, which would be bad for both of us."

"Yes."

"Of course, what I really want is for you to forget school and take over the agency. But, if you go part time, that's okay."

Steve's priorities are shown below in Figure 6.

Steve's walk-away strategy is to give Lisa the business and his walk-away value is "pretty bad." This can be seen by examining Figure 6. He knows that Lisa won't forget about school, and he rejects his options of favoring, ignoring, or putting his foot down. This leaves only offering her the business or being deliberately nasty. If he's nasty, the result is "bad." If he offers her the business, the worst is that she turns it down to go to school full time. This is "pretty bad," which is better than "bad."

Notice that both have tried to be completely honest with each other. This is like a labor-management negotiation, in which the company opens the books for the inspection of the union. In the case of Steve and Lisa, both are opening the books because they both want to fully understand each other's priorities.

What does game theory tell them to do? There are various solutions, depending on which game theoretician one pays attention to. This chapter is based on a procedure developed by Lloyd Shapley. As this book is nonmathematical, we shall look only at the ideas underlying the computations and skip the mathematical nuts and bolts.*

To be satisfied with a deal, people should feel that they do better with it than without it. Steve's walk-away value is "pretty bad" and Lisa's is "not very good." So, whatever the final agreement, each should do better than his or her walk-away values. From Figure 5, Lisa sees four possible outcomes which are better for her than her walk-away value. However, she can immediately throw away two of them because they require that Steve favor school, which she knows only too well he's not about to do.

To find the best compromise, she simply compares

*For the technically inclined reader, this discussion assumes that the rankings of Figures 5 and 6 can be represented by consecutive integers.

AWFUL!
I favor or ignore/Lisa goes full or part time
WISHFUL THINKING
GREAT!
I ignore/Lisa forgets school
I'D FEEL LIKE A JERK
I'm nasty/Lisa drops school
BAD
I'm nasty/Lisa goes full or part time
PRETTY BAD
I give her the business /Lisa goes full time
OK!
I give her the business /Lisa goes part time
WHAT I REALLY WANT
I give her the business /Lisa forgets school
DOESN'T WORK
I put foot down/Lisa does anything

Figure 6

the two remaining outcomes with Steve's priorities to see which is best.

Suppose Lisa goes to school full time and Steve offers her the business. For Lisa this is "good," much better than her walk-away value. However, for Steve, this is "pretty bad," his walk-away value—not good enough for a compromise. What's left? Only the outcome where Lisa goes to school part time, and Steve gives her the business. For Steve, this is "okay," better than his walk-away value. For Lisa, this is "pretty good," also better than her walk-away value. This *has* to be it! There is no other mutually agreeable compromise—we've looked at all the others.

Game theory shows who should make the major concession and who has to be favored by the compromise.

There are other game-theory techniques involving threats which could be used in this situation, but Steve and Lisa have tried to avoid threats, believing that once people start threatening, friendship gets lost. So they have based the compromise completely on a comparison between their walk-away values and the possible agreements. Neither wants to walk away and neither has to, because they both are better off with this solution than with no solution.

Notice that in the compromise, Steve uses his walk-away strategy, giving Lisa the business, while Lisa doesn't use hers. The compromise requires that she use a different option altogether. She could, of course, exploit the fact that Steve is using his walk-away strategy—she could go to school full time. But, she doesn't. Why not? She wants a permanent solution, about which both feel good.

Steve may be idiosyncratic, but Lisa is no fool. Nor is she self-sacrificing. If all Steve gets out of the deal is what he could get without the deal, he may question other parts of their relationship. True, this may result in more concern for her feelings. Alternatively, his self-questioning may produce a search for someone who "understands" him, and who wants his kind of attitudes. Lisa doesn't want to lose him.

In any compromise, nobody wins everything. A glance at Figures 5 and 6 shows that this solution is not *the* best outcome for either of them. It is, instead, the most satisfactory outcome for both of them at the same time, which is what making deals is all about. If you want to have it exactly your own way you should be prepared to go your own way.

Having worked out what she feels she should do, Lisa says: "Steve, if I'm just considering me, the best thing is to go to school full time, get it over with and still have the option of choosing to run the ad agency. But, when I consider everything—my relationship with you and what I want from our marriage—then I think the best thing for all concerned might be for me to go to school part time and take over the agency."

"Great!"

Should Lisa have told Steve to go to hell, and then done what she felt was best for herself alone? Feminists may say yes. But Lisa is not a feminist. She has spent eighteen years of her life married to a man who has been the dominant partner in their relationship. She doesn't want to divorce him or spoil her domestic life. Rather, she wants a settlement within the present context, which can only be achieved by compromise. She realizes, in short, that she *must* take his feelings into account, even if they are partially irrational and idiosyncratic. For all his faults, she loves her husband. As she put it: "Eighteen years of a happy marriage has been a reality, architecture has only been a fantasy."

* * *

This chapter has shown how to find a possible compromise, if there is one.

FIGHT

First, don't be afraid to begin with a fight, which can bring important, but hidden, feelings out in the open. Steve and Lisa began the process of making a deal with a fight.

OPTIONS

Second, figure out your options. Often, the best way to do this is by talking the matter over with a friend. Be careful not to take any advice as to how you should feel or act. You know how you feel, and *your* actions should be based on *your* feelings. But, listen carefully to advice on what options may exist. Both Steve and Lisa heard of options they hadn't previously considered as a result of talking things over with friends. Try to get the person with whom you have the dispute to talk things over with someone.

PRIORITIES

Third, make a preliminary assessment of your priorities. Obviously bad choices should be recognized as such. Try to get the person with whom you want to settle things to do the same.

OPEN THE BOOKS

Fourth, talk over your priorities with each other. Each should be fully aware of how the other sees things. At this point, each should work out a walk-away strategy. If necessary, break off talking for a while so this can be done. Labor-management negotiators frequently break so that each side can assess its position.

WALK-AWAY VALUES

Fifth, make sure you are aware of the other's walk-away value, and that he or she is aware of yours.

COMPROMISE

Sixth, look at each possible compromise. See if it is better than your walk-away value and better than that of the other person. If not, forget it and look at another possible settlement. A compromise which is better than one person's walk-away value, but only as good as the other's, isn't good enough. The one with the short end is bound to ask, "What kind of a deal is this?" If there is more than one possible compromise which fits the bill, pick the one which is best for both. If there are several of these, some favoring one person, some the other, we'll take up the matter in Chapter 8, "Let's Be Civilized."

6

Goodbye, Charlie

The first action of the new Communist government of Cambodia was to cut off communication with the outside world. Foreign embassies were told to pack up and get out. Foreign news personnel were given the boot. The only photographs initially to come out of Cambodia were taken from across the river in Thailand.

Apparently, the new regime feared that any foreigner was a potential trouble-maker, or CIA agent. The foreign countries had no leverage, since the new government wanted nothing from them. So all they could do was pick up their marbles and go home.

A similar situation can happen at a trendy night spot. Not everyone's money is good enough, as turned-away night owls can testify. The club, of course, feels that it doesn't need anyone beyond the already acceptable.

Although rejection may be common in interpersonal

relations, it is usually cushioned by an attempt to avoid hard feelings. Occasionally, however, the door is slammed in one's face. The question is, can the rejector be sure of the consequences?

Jane, thirty-four years old, had been married for seven years, but has been divorced for the past three. She is now the assistant fashion editor for a medium-sized magazine.

Due to an internal disorder, she has been medically diagnosed as incapable of conceiving children. Resigned to this fact of life, Jane has thought birth control to be unnecessary. As do many single people her age, she occasionally gets involved in casual affairs.

One night she sleeps with Charlie, whom she knows through the magazine. Two weeks later, Jane is surprised to miss a period. Her doctor surprises her even more: "You're pregnant."

How did her disability suddenly clear up? Nobody knows. However, knowing that she is unmarried, the doctor offers to arrange for an abortion.

"If I have an abortion, could I become pregnant again?"

"There are no guarantees."

Not knowing what to do, Jane tells the doctor she'll think it over. She establishes her options and priorities in a conversation with her friend Sue:

"What really worries me is having an abortion and then afterwards, the doctor tells me 'that was it'—I'd have missed my only chance to have a kid, which would be terrible."

"Yes."

"Of course, it would be best not to have this baby, but

be able to have children by a man of my own choosing."

"Jane, what about Charlie?"

"Charlie? What about him?"

"Don't you think you should tell him?"

"What for?"

"Well, if you decide to have it, child support! I mean, it is his kid."

"I don't want anything from him. He'd have a claim on me. And I see absolutely no reason to give him one. It's almost as bad as being childless."

"He may want to visit the child."

"I doubt it, but suppose I have this child and then meet a man I really care about. I don't want to set any precedent for Charlie coming around."

"Yes, that would be bad."

"Well, pretty bad, I could always tell him to get lost."

"You might get to like him."

"I doubt that."

"So you're not going to let him visit?"

"No. It's much better to have the kid and forget about Charlie. I'd have my own life which includes a child, period. I'm not going to tell Charlie anything about it."

Jane has worked out her priorities, see Figure 7.

The worst that can happen if Jane has the baby but ignores Charlie is "okay." The worst with any other choice isn't as good, so her walk-away strategy is clear —have the baby and forget Charlie.

Nine Months Later: Through mutual friends at the magazine, Charlie finds out about Jane's new son, and realizes that he could well be the father. So, he phones: "Is it?"

Figure 7

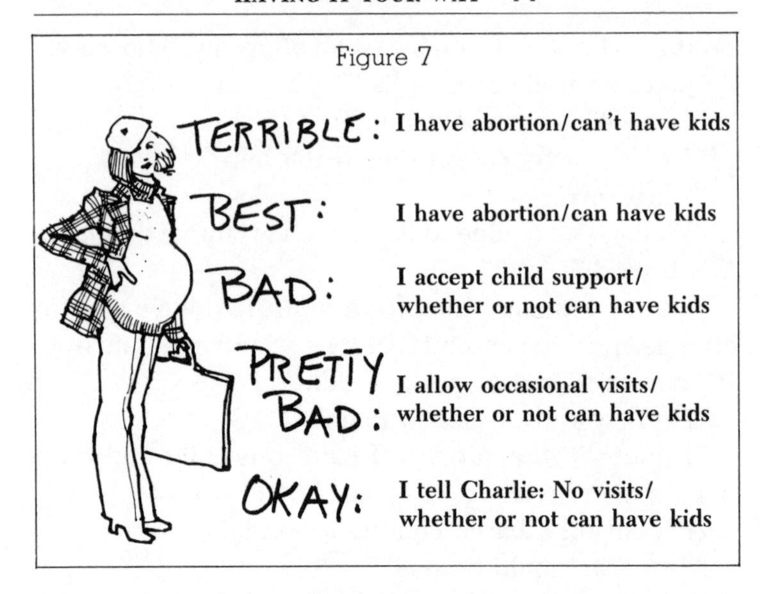

TERRIBLE: I have abortion/can't have kids
BEST: I have abortion/can have kids
BAD: I accept child support/ whether or not can have kids
PRETTY BAD: I allow occasional visits/ whether or not can have kids
OKAY: I tell Charlie: No visits/ whether or not can have kids

"Yes."

"That's what I thought. Our son will eventually want to know his father. Since I'll probably have some dealings with him anyway, I'd like to establish them now."

"Our son? Don't be silly, you're out of the picture. I don't see why you want to have anything more to do with it."

"What do you mean? I'm the father."

"Technically, yes. Emotionally, no. We went to bed for fun one night. It wasn't much fun, but I got pregnant anyway. At the time I had every reason to believe that I couldn't. But for some reason, I did. I had the child only because I believed that this was my one chance. I hate to be so blunt, but had I known I was capable of getting pregnant, I'd have found someone more suitable."

"I think you're being callous."

"I don't mean to be, just honest. A woman who can get pregnant whenever she wants can pick her man. I had to take pot luck. If I had a real option, I wouldn't have a moral leg to stand on. But, as it is, I'm right, and you're ridiculous."

"Look, it's just that I'd like to be able to see my son."

"Most men would be happy to avoid the responsibility."

"I'm me, not most men. As it happens, this is my only child too, and I want to see him. I'd be willing to pay child support."

"Charlie, I think it's sweet of you to offer, but I don't want your money, and I don't want you in my life. Our scene—if you can call it that—is over; it was just a casual one-night stand."

"If you don't want child support, so much the better. But I'm willing to pay if I can visit. It's a standing offer."

"Charlie, my situation is very simple. Your visiting the baby, with or without child support, is out of the question."

"Well, I've offered."

"Goodbye, Charlie."

Jane genuinely feels that Charlie has no valid moral claim, and she's not interested in his wishes, needs, or offers. She doesn't want a "fair" solution, a compromise that would recognize Charlie's claims and allow him to visit on a regular basis. So negotiation is irrelevant.

However, Jane is not heartless: "If the situation were

reversed, I'd be at least curious as to what my baby looked like. Maybe I should let Charlie come over once to see the baby. Just once."

She phones Charlie, inviting him over the following Sunday, between five and seven in the afternoon.

The outcome was nonnegotiated, and based strictly on the relative power of each. Jane chose "Goodbye, Charlie," and Charlie offered to pay. With this combination of choices and no changes of heart, neither would budge from his or her position. For example, if Charlie withdraws his offer of support, and just asks to visit, he wouldn't feel as good about himself. Similarly, if Jane changes her strategy and lets Charlie visit even on an occasional basis, she feels worse. So, both are locked into their choices.

* * *

Postscript: After six weeks away from work, Jane returns. The senior editors of her magazine, who have previously been sending her to Paris and other glamour spots, now refuse to do so: "Jane, a young baby should be with its mother, and shouldn't be packed around the world like baggage. You'll be based here in town for the next couple of years."

When Jane protests, they add: "The welfare of a child is nonnegotiable. The subject is closed."

Later she finds out that Charlie and one of the senior editors are bridge partners.

* * *

This chapter has shown that it takes two to bargain. If you refuse to do so, you may be able to get away with it, but make sure your flanks are covered. Punishing strategies are about holding people in fixed positions and two can sometimes play at that.

The best thing is usually to keep the dialogue open, at least until you thoroughly understand the consequences of shutting it off.

7

Don't Take, Have Him Give

Recently, the papers have carried stories about cities, even countries, going broke. Zaire, the former Belgian Congo, has gone bankrupt. New York City continues to struggle. New York State remains on the edge, a problem brought about by the city's difficulties.

When the city was teetering, people who owned its bonds wondered how they could get their money back. One bondholder jokingly talked of taking over a block of Forty-third Street, and forcing everyone who passed to pay tribute. The point is, if the city goes broke, there is no way to make it pay up.

Recovering private debt, however, is a cut and dried procedure. Contracts and courts say who should pay what, and when. But many loans are informal—one friend helping out another. If repayed on time, great. But what do you do if the borrower doesn't come through when he's supposed to? This chapter shows how to deal with the problem and still maintain the friendship.

* * *

Although never lovers, Ann and Joe have been friends for years. Joe is an independent businessman, Ann a social worker. He's generally flusher than she is, and several times he's bailed her out of tight situations. Ann never failed to repay these friendly loans on time. Recently, however, Joe wanted to conclude a deal on a new condominium, but was short a few thousand. Not wanting to pay interest, and feeling no qualms about borrowing from someone he has helped out in the past, Joe asked Ann for a friendly loan. She had the cash, and was only too happy to lend it. The understanding: Joe would return the money in three months, by which time Ann expected to need it.

Four months pass—no money from Joe, but now Ann needs at least part of it. She jokingly alludes to her tight situation but he doesn't take the hint. She figures that although he may not have the cash, he could get it. His sensitivity to her needs has thus become an issue.

Completely confused about what to do, she talks to her friend, Alan: "I need a third of the money just to meet current bills. However, if I really press the matter, even if I get the money, I'm afraid I might insult Joe, losing an old friend."

"Oh, I don't think so."

"He seems to be avoiding me, so I'm wondering how he really feels. I mean, I lent him my money. Why should the burden of getting it back fall on me, too?"

"He's probably having a rough time."

"But he should say something."

"You could say something."

"Well, I haven't wanted to, but if I do ask him outright for the money, the question is how?"

"Obviously, you can phone or go and see him."

"Or ask someone else, like Leo, to talk to him. I've even thought about bringing it up in public, to embarrass him into paying. That's how frantic and crazy I'm getting."

"Why not write him?"

"No, I never write letters."

Ann sees four possible responses by Joe: Welsh or cough up the cash and either way, he could be happy about it or not. All the possible combinations of Joe's responses to each of Ann's options are shown below in Figure 8.

Ann, with a large list of priorities to sort, will begin as usual by looking at what she doesn't want. The list is so long that she cannot trust to luck, but must deliberately ask herself how she feels about each outcome. This is important because her problem should be handled with subtlety, and this requires an understanding of the intricacies of her own feelings.

"Ann, are you seriously considering making a public scene?"

"No, but because the thought has crossed my mind, I figured I better bring it out in the open. It would be the worst thing I could do; it would demean me."

Ann writes "worst" in front of every result of "public scene."

Although Ann never really intended to make a scene,

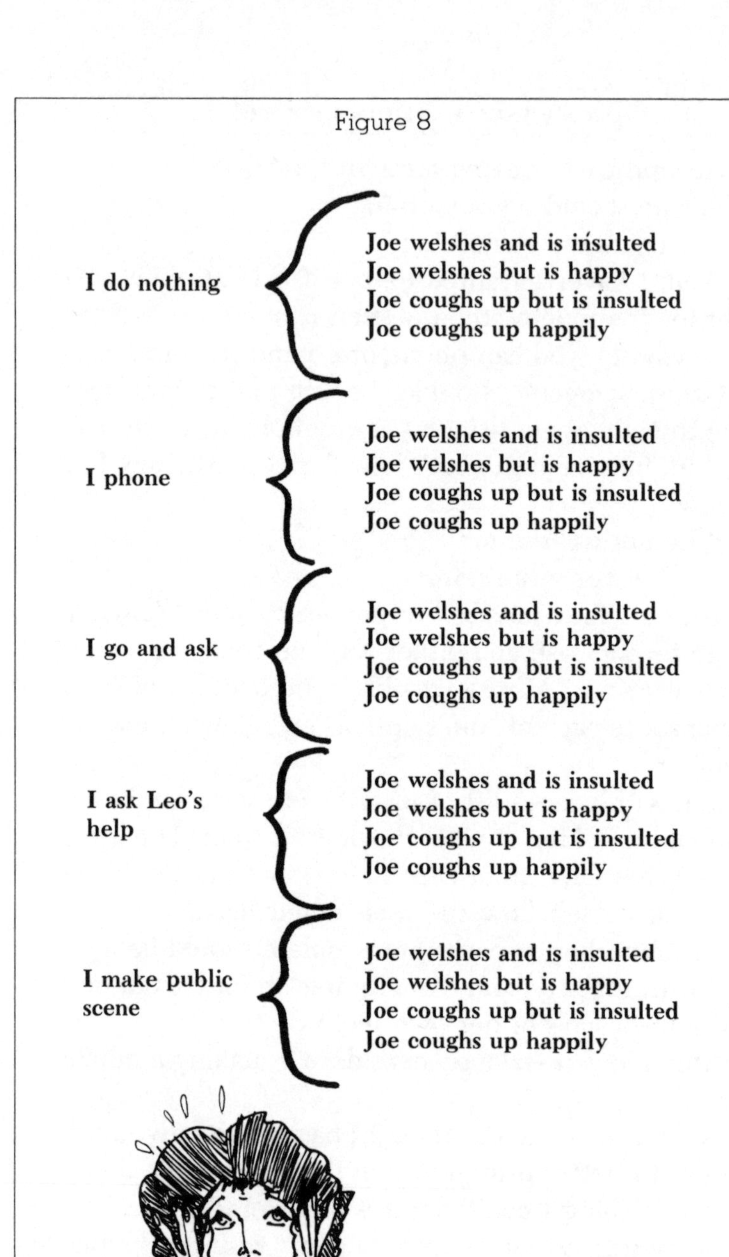
I do nothing
Joe welshes and is insulted
Joe welshes but is happy
Joe coughs up but is insulted
Joe coughs up happily
I phone
Joe welshes and is insulted
Joe welshes but is happy
Joe coughs up but is insulted
Joe coughs up happily
I go and ask
Joe welshes and is insulted
Joe welshes but is happy
Joe coughs up but is insulted
Joe coughs up happily
I ask Leo's help
Joe welshes and is insulted
Joe welshes but is happy
Joe coughs up but is insulted
Joe coughs up happily
I make public scene
Joe welshes and is insulted
Joe welshes but is happy
Joe coughs up but is insulted
Joe coughs up happily

Figure 8

she is right to put this option on her list, where she can see it starkly contrasted with more sober ones. Otherwise, it would always play at the back of her mind, as if she had a hidden power which, in fact, she doesn't have.

"Just as bad would be if Leo talked to Joe, who then got insulted and didn't pay."

"That would be about as bad as you could get—you don't get your money, you embarrass yourself by bringing in somebody else, and you lose a friend."

"Yes, a triple play."

"Do you have a written record or signed statement of the loan?"

"Of course not. It's backed by our friendship, which has always meant a lot to me."

"Well that eliminates embarrassing him in any deliberate way."

"Yeah, I guess I don't really want to bring in Leo, no matter what."

Ann writes "bad" in front of the remaining "ask Leo" outcomes. Once again, by bringing an obviously undesirable option out in the open, she is able to throw it away, and not secretly dwell on it.

"But Alan, I have a nagging, and I'm sure, irrational fear that if I do nothing, not only won't I get my money, but I'll have lost a friend, too. Sometimes you really don't understand people, even those you've known for years. He had always been the one with money. Maybe now he resents me because I had it when he didn't."

"I don't think so, but let's worry about your feelings now, not his."

"Wait, this is important. Maybe he's already offended. It's ridiculous but possible."

"Well, do you question your friendship because he's making himself scarce?"

"Sure that raises doubts about the friendship, but at least *I* haven't done anything stupid to provoke him, so it's not the worst, but it's very bad.

"I think it's just your paranoia, Ann, but put it down."

Ann writes "very bad" in front of "do nothing/Joe welshes, and is insulted."

"Phoning him wouldn't be stupid."

"Yes, and even if I don't get my money and I do lose a friend, I mean, it's obviously bad, but the weird thing is I'd prefer doing something to doing nothing. Sometimes, resolving doubt is good for its own sake. At least, I'd find out about Joe's craziness."

"Then asking in person would be even better."

"Well, if the results are as in the last case, this would still be pretty bad."

Ann can now establish her walk-away value. She realizes that none of the remaining outcomes are worse than those in which she makes a fool of herself, or in which she fails to get her money but succeeds in losing a friend. These are the ones she's already considered, and the best of them is "pretty bad": i.e., she goes to Joe for her money, but he gets insulted and doesn't pay up. Going in person is her walk-away strategy, and, in this case, she would walk away with empty pockets.

Often, when you have a long list of outcomes to evaluate, the best approach is to quickly pick out the obviously bad ones. Do so for each of your options and you get your walk-away value at once. This will make the rest of the evaluation easier. Ann now has a base mark by which to measure the value of the remaining outcomes.

The question which she still hasn't worked out is how to get her money back and keep a friend. Game theory will tell her how, providing she examines the rest of the outcomes. The best approach will then be clear, including what she should actually say to Joe.

"I don't think he'll refuse you or get insulted."

"Well, it would be best to do nothing and have Joe pay what I need without any resentment."

"Naturally."

"Of course, if he pays up, even if for some reason he's angry, at least I'd have my money. That'd be okay with me if that's how Joe would want it."

"Suppose he isn't resentful but doesn't pay?"

"Well, Alan, this is where I am now, and it's not so good."

"Of course, you're not helped out at all."

"But, suppose I press the point."

"And he doesn't come through?"

"Yes, but still acts like a friend. I'd really find out then, wouldn't I? That would be pretty bad."

"Yes, a good, unreliable friend. I've got lots of them. Of course, he may pay up, but get mad about it."

"Well, that's what worries me. It wouldn't be so good. I'd have asked in a reasonable way for what's mine, and have lost a friend over it—I guess if that were the case, he wouldn't have been much of a friend after all."

"He may not have the money, having to suddenly get it may annoy him."

"True."

"The smoothest way to get it would be to work out something between the two of you."

"Right."

"And that's hard to do over the phone."

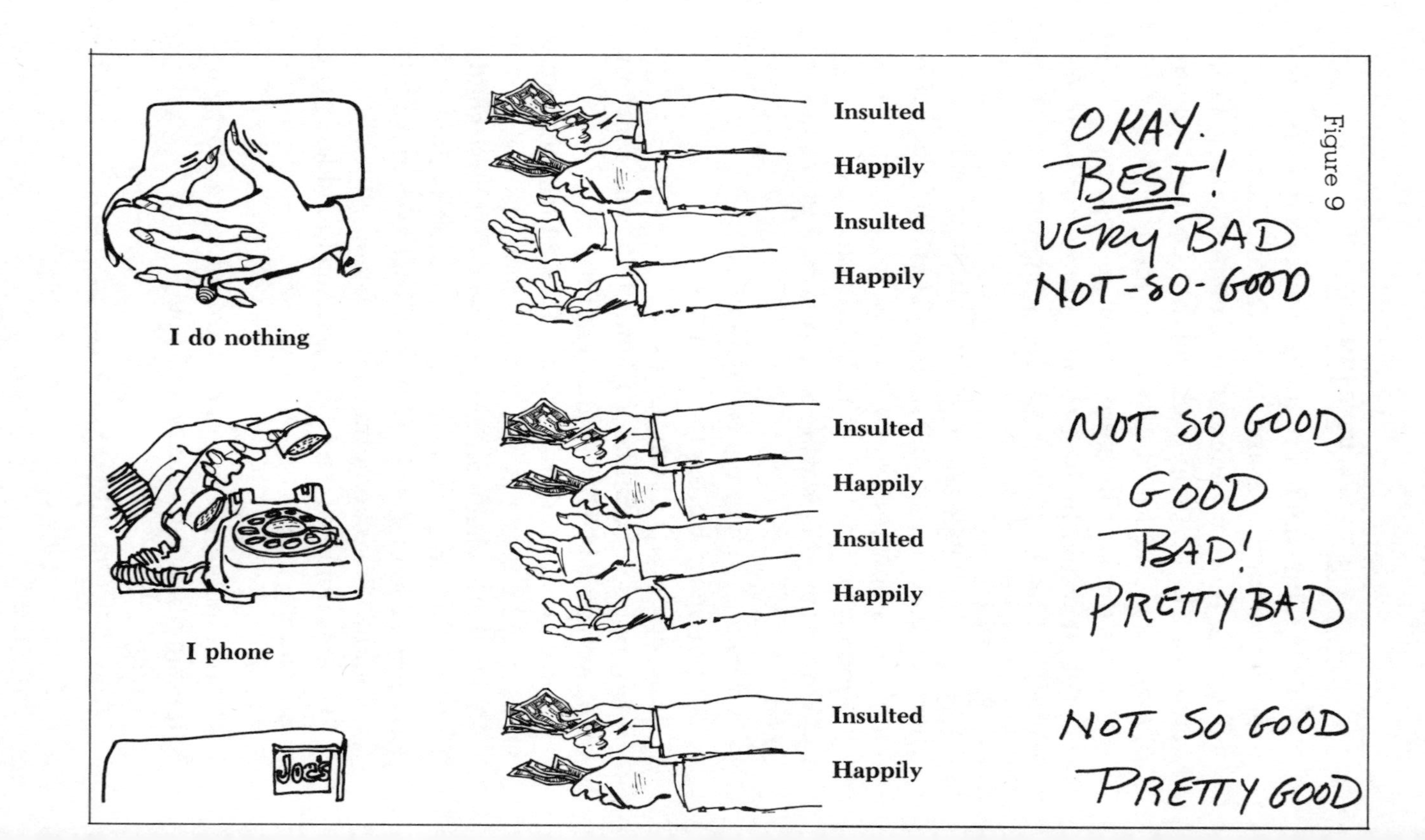
I do nothing
Insulted
Happily
Insulted
Happily
OKAY.
BEST!
VERY BAD
NOT-SO-GOOD
I phone
Insulted
Happily
Insulted
Happily
NOT SO GOOD
GOOD
BAD!
PRETTY BAD
Joe's
Insulted
Happily
NOT SO GOOD
PRETTY GOOD

Figure 9

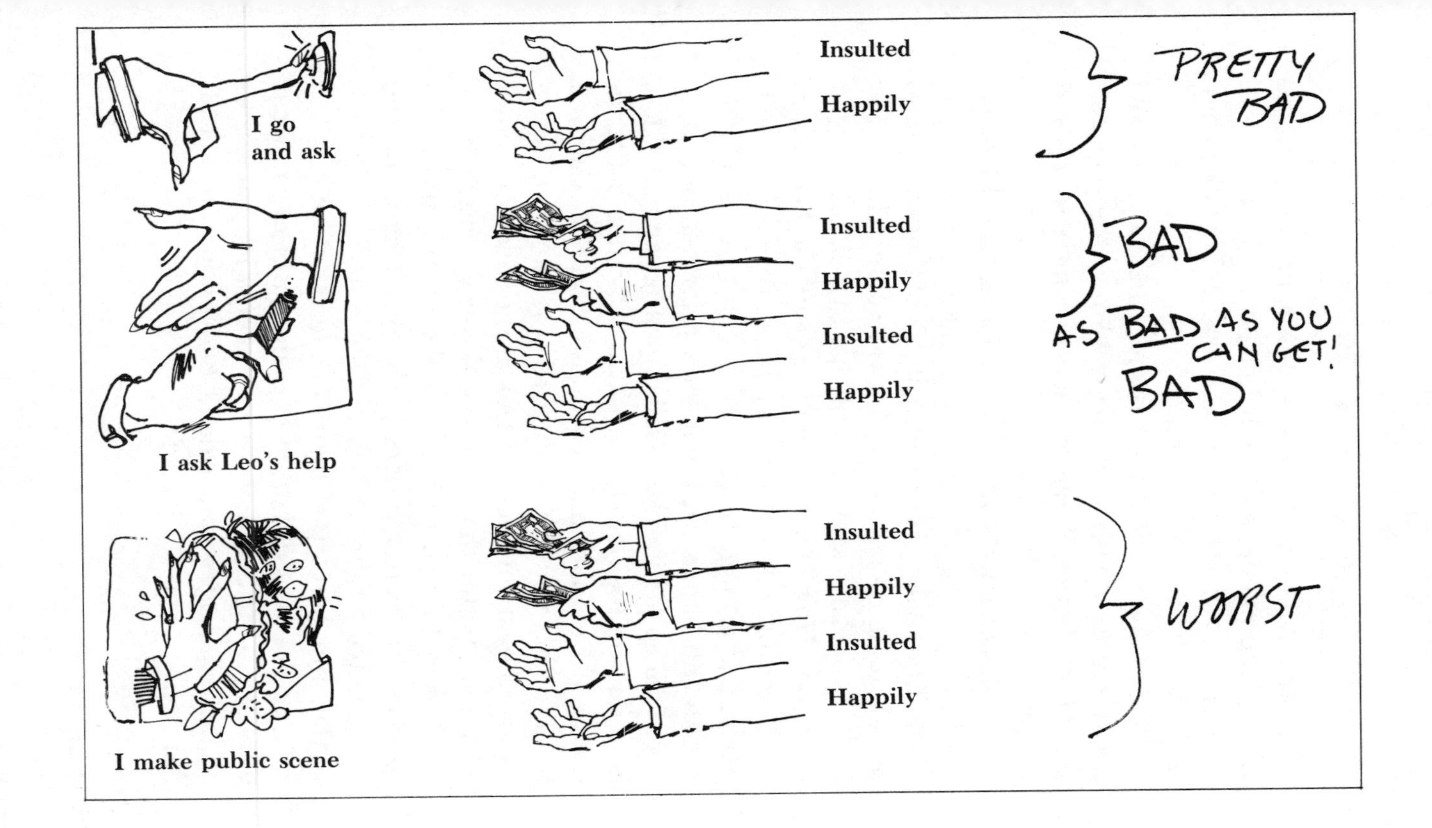
I go
and ask
Insulted
Happily
PRETTY BAD
I ask Leo's help
Insulted
Happily
BAD
Insulted
Happily
AS BAD AS YOU CAN GET! BAD
I make public scene
Insulted
Happily
Insulted
Happily
WORST

"If we have to work out something, he could tell me that on the phone, and then I could see him."

"But don't you think it's more graceful to talk to him in person from the beginning?"

"Not really, we've always been casual. So, if I could get the money with a 'by the way' over the phone, it *would* be better. I mean, on anything else I'd just phone."

Ann has now worked out her priorities, see Figure 9, pp. 86–87.

Did Ann really have to evaluate all of these outcomes merely to figure out her walk-away value? The answer is no. As we have seen, she only has to look at the worst outcomes. However, for *making deals* she has to look at all the others—or she may be surprised by a self-made, bad arrangement.

Ann still has to work out Joe's priorities. She can't very well go to him like a Gallup pollster and ask, "Joe, how do you feel about not paying and not being insulted, if I do nothing?" Obviously, she and Alan will have to figure out his feelings as best they can. Fortunately, they'll only have to look at the seven outcomes which are better than Ann's walk-away value. She doesn't want to deal with any of the others, and if she handles the matter properly, she won't have to. The seven outcomes are shown in Figure 10.

Alan and Ann try to figure out Joe's priorities:

"Ann, Joe isn't dishonest or malicious. I mean, what we were doing before was good therapy for you—exorcising your anger—but I don't know how relevant all the options were to Joe."

"Yes."

Figure 10

NOT SO GOOD	I do nothing	/Joe welshes but is happy
OKAY	I do nothing	/Joe coughs up but is insulted
BEST!	I do nothing	/Joe coughs up happily
NOT SO GOOD	I phone	/Joe coughs up but is insulted
GOOD	I phone	/Joe coughs up happily
NOT SO GOOD	I go and ask	/Joe coughs up but is insulted
PRETTY GOOD	I go and ask	/Joe coughs up happily

"The whole basis of your relationship is that he's a good friend and ultimately other people are involved as well."

"There aren't other people involved."

"Yeah, but the situation could come up: What will other people think?"

"It's possible. I don't think anybody but you knows at the moment."

"It seems to me that what he really wants now is not to pay and hope everything stays cool."

"I'm pretty sure he doesn't have the cash, or he'd have given it to me. To get it, he'd have to sell something, or borrow from someone else. I know he has assets he can sell, but obviously he doesn't want to."

"He might, if the circumstances were clear that he was helping you."

"Yes, but it's not something he *wants* to do. He'd prefer to maintain the friendship and not sell anything."

"Sure, but doing what's necessary is probably okay for him."

"Providing I treat him right."

"Sure, if you brought in Leo, or made a public scene, it would be a lot harder on him. He'd pay, but feel bad that he allowed things to get to the point where you were desperate."

"I think so."

"And he might think less of you for not having confronted him earlier, feeling that you're weak or that you have a very low regard for the friendship."

"Yes, of course those items are off my list, anyway."

"Right, but they probably establish his walk-away value."

Unless at least one item on the list is better for Joe than his walk-away value, he may not want to make a deal, and Ann can probably kiss both her money and her "friend" goodbye. So there's no harm in guessing his walk-away value, even though it's off the list.

"Suppose I do nothing, and he pays, but resents it. . . . He'd ultimately feel like a dope when he realized that he'd lost a friend for reasons known only to himself."

"Well, that would be *his* craziness. But if you talk it out, he still might get angry because you're forcing him to do something."

"I don't think he'd stay mad for long."

"Probably not."

"The question is, would he rather pay up happily or be mad about it?"

"I don't know. Do you know how he thinks?"

"Not really. Let's figure it as okay either way."

"That's about as refined as we can make it. Except if he pays, he's morally off the hook."

"Yeah, but even so, there's no telling how he'd feel about being pushed."

Although aware that much subtlety has been left out, Ann and Alan have now established what they think are Joe's priorities, see Figure 11.

They've also guessed his presumed walk-away value, "bad." He gets this if he pays up when Ann brings in Leo or makes a public scene.

What will the final arrangement be like? The outcomes in Figure 11 are the only ones which are better than the walk-away values of both Ann and Joe. Presumably both would want to do better than their walk-away values, so all of these are potential bargains.

Obviously, some of them are of no help to Ann, for example, the third one on the list. Ann does nothing and Joe pays up happily. This is "best" for Ann and "okay" for Joe. However, there is no way for Ann to inform Joe of this happy outcome without doing something, which means using one of her other options!

"How long will you give him on this outcome?"

"Until the end of this sentence, Alan."

Figure 11

Ann's Priorities			Joe's *Presumed* Priorities
NOT SO GOOD	**I do nothing** /	**Joe welshes happily**	WHAT HE *MUST* REALLY WANT.
OKAY	**I do nothing** /	**Joe coughs up, insulted**	NOT GOOD – HE'D FEEL LIKE A DOPE
BEST!	**I do nothing** /	**Joe coughs up happily**	
NOT SO GOOD	**I phone** /	**Joe coughs up, insulted**	
GOOD	**I phone** /	**Joe coughs up happily**	PROBABLY OKAY – BUT WE DON'T KNOW EXACTLY HOW HE THINKS.
NOT SO GOOD	**I go and ask** /	**Joe coughs up, insulted**	
PRETTY GOOD	**I go and ask** /	**Joe coughs up happily**	

Ann is currently getting an outcome which is "not so good." She is doing nothing and Joe isn't paying up, but seems to be happy about it. Ann can do better by either phoning or going to see him in person. Either course will slightly reduce Joe's happiness, but that's his tough luck.

Which should Ann do, phone or go in person? If she phones, she'll get an outcome which is either "good" or "not so good," depending on whether Joe is happy or offended when he pays up. On the other hand, if she goes in person, she does not do as well. She'll get an outcome which is either "pretty good" or "not so good," again depending on whether or not Joe gets angry when he pays.

Ann's bargaining strategy is clear: She should phone and politely ask for the money. If that doesn't work, she should make the point face to face.

Suppose Ann phones. The outcome where Joe pays, but is insulted is "not so good" for Ann, but "okay" for Joe. On the other hand, if Joe isn't insulted, the result is still only "okay" for himself, but "good" for Ann. Unless Joe is particularly perverse, he'd choose this outcome, since Ann is better off than in the previous case and Joe is no worse off. In short, being friendly and generous costs Joe nothing more, if he's going to pay up anyway. This is a standard piece of traditional wisdom in bargaining. We have just shown its soundness through game theory. Generosity turns out to be nothing more than a special case of the principle of both people doing as well as possible.

What should she actually say to him? Again, this is in the game theory. She's worked out both walk-away values—hers and his. She's seen how much both benefit by his paying, even if he has to sell something to raise the cash. The friendship is maintained. He's better off than he would be if he doesn't pay, going from his walk-away value of "bad" to "okay." But she goes from "pretty bad" to "good," a major gain. Although both are better off, the gain for her is decisive. She should literally tell him that! And, she can also point out that this outcome is not the very best for herself, which would have been to get the money without any hassle at all. So, she isn't getting everything, and neither is he. Most important, she should make sure he knows that she would feel bad if he gets offended. The solution shows that the best way to deal with the problem is not in terms of individual but of mutual gain. In other words, she shouldn't take, she should have him give. Both should feel ahead.

What about the actual words she should use? Books on bargaining sometimes contain examples of set speeches, to be used in the appropriate circumstances. They are always hedged with a warning that the speeches are merely the drift of what should be said, and should be put in the bargainers' own words. This book contains no such speeches. Ann has known Joe for years, and knows how to talk to him. Her problem concerned what she should say, not how she should say it.

Unless you have Richard ("Ask not what your country can do for you . . .") Goodwin or a comparable speechwriter on tap, do the best you can in your own words. Stock phrases should be kept in stock, and not put on display.

Ann talked to Joe, who sold an asset to pay her. They're still friends.

* * *

This chapter has shown how to get a friend to pay back the money you lent him.

IRRATIONAL OPTIONS

First, work out your options. Include seemingly irrational ones, such as making a public scene. These should be brought out in the open so that you don't secretly dwell on them.

BORROWER'S OPTIONS

Second, work out your friend's options. Include the possibility that he might take offense.

WALK-AWAY

Third, establish your walk-away value as early as possible. You will probably have a long list of outcomes to examine. Finding your walk-away value early will help in the analysis.

COMPLETE LIST

Fourth, complete the evaluation of all the outcomes.

GOOD DEAL

Fifth, make a list of those outcomes which are *better* than your walk-away value. You hope to get a good outcome from this list.

BORROWER'S PRIORITIES

Sixth, using only the outcomes on the list made in step five, try to figure out your friend's priorities. Also, try to make an intelligent guess concerning his or her walk-away value. If none of the outcomes on this list are better than your friend's walk-away value, you can probably kiss your money and your "friend" goodbye.

BEST DEALS

Seventh, work out the best realizable outcome for both of you from this list, as explained in the text.

TALK

Eighth, talk to your friend, and explain why he or she should pay up, giving as your reason the criteria for selecting the outcome chosen in step seven. Do not dwell on the morality of the issue. This is in no way part of the game theory, and your friend's morality may turn out to be surprisingly different from your own.

8

Let's Be Civilized

The claim that half of all marriages made today will end in divorce is pure hokum. Here's how it's cooked up: Suppose that in your town this week there are one thousand marriages and five hundred divorces. A 50 percent divorce rate? That's what the phoney claim says. But the people getting divorced probably aren't the same ones who just got married. This week there may be one thousand deaths of married people. If we use the same reasoning, marriage is lethal—a 100 percent death rate within one week.

The actual divorce rate in this country has been slowly increasing and is soberly estimated to be not 50 percent, but about 2 percent a year. In other words, of all marriages existing in a given year, roughly 2 percent end in divorce *that* year. Over a few years this adds up and we begin to notice a lot of divorced people. But if we look further, we see far more who stay married.

Whatever the statistics, for those who go through a divorce, the experience is almost always wrenching, and often guilt-ridden, but rarely the courtroom drama shown on TV. Indeed, the fact that divorce is shown on TV as a courtroom battle is evidence of how little most people know about it. Most divorces aren't fought out in complex litigation. Most, in fact, are fully settled before they get to court, the actual court hearing usually lasting five minutes, with nobody struggling—except perhaps to stay awake. The real conflict, if there is any, has already taken place in the lawyers' offices—over the details of a separation agreement, which the court will most likely rubber-stamp. Often, however, the negotiations for these agreements drag on for months, because one spouse holds out for more, or changes his or her mind after a tentative agreement has been hammered out.

This chapter shows how to deal with someone who repudiates an agreement. Although the chapter focuses on a specific divorce, many of the ideas are applicable to the repudiation of any agreement among former friends.

Everybody knows that the law makes getting married much easier than getting divorced. But the discrepancy would be true even if the law weren't rigged in this way. Why? When two people marry they both want happiness and believe they can get it from each other. But when they divorce they know this isn't the case. Their differences seem greater than their common interests. So, if they make a tentative separation agreement, what is there to keep one person from having second thoughts, repudiating the agreement, and asking for more? Certainly not fear of displeasing the other one—on the contrary, there may be pleasure in making the other one miserable. Despite this possible attitude, which did not exist in any of our earlier examples, mutual agreement is not only possible, but happens all the time. However, it must include the consideration of mutual punishment. This chapter will show how to figure a different kind of walk-away value, really a punishing value. Based on this type of walk-away strategy, we shall show how to cooperate—the cooperation of two people with daggers hidden under their cloaks.

* * *

After seven years of marriage, Rick and Joyce divorce. They had been terrifically romantic for six months before their marriage, but things cooled down soon afterwards. They began to get on each other's nerves. Rick found that many close friends from bachelor days were no longer close, he thought, because of the marriage. Very social, he was upset by the loss. Joyce objected to Rick's financial situation. He was in the antique business, an economic roller coaster—ten thousand in the checking account one month, nothing the next. During the slack periods she carried them both. She was a teacher in the public schools, with a regular income, and would have liked Rick to get a steady job. She preferred that he earn a nominal amount every week to gambling on fifty thousand by the end of the year. Rick believed that her feelings were neurotic, an overreaction to her wheeler-dealer father. He had been very rich at one point, very broke at another.

Finally, after six months of hinting around, Joyce proposed a divorce. Rick's attitude: "If that's the way you feel, you can leave. Get the divorce, I won't fight it."

Finally, however, Rick was the one who moved out, feeling: "This is no way to live my life." They talked of a trial separation, but it was obviously permanent.

Rick moved into an apartment, temporarily leaving Joyce with the house, all the personal effects, the household furnishings, and a large number of antiques. Joyce quickly tired of seeing Rick drop by to pick up this or that and, after a few weeks, changed the locks, not giving him a new set of keys. At first outraged, he eventually calmed down, deciding, "Let's be civilized."

The house, having been acquired during the marriage, was jointly owned. But the major furnishings and antiques belonged largely to him, 80 percent of them having been bought before they married.

They went to Rick's lawyer together, with the aim of saving money. However, the lawyer insisted that Joyce get her own attorney to represent her needs and her interests. She protested, but eventually got one through an old friend.

Rick and his lawyer suggested that Joyce and her lawyer draw up a divorce agreement and submit it to them. This she did. Unfortunately, Joyce misunderstood certain details of property and debt. Rick and his lawyer made the corrections and signed the agreement. Although not happy with the changes, Joyce signed, her lawyer advising: "These corrections are fully documented; the agreement is a straight down the middle division of community property. If anything, it favors you."

But, the alterations annoyed her. She had supported Rick during several lean years, saving him from selling any of his reserve antiques. Nevertheless, Joyce's generosity gave her no *legal* claim to his property.

The alterations continued to eat away at Joyce and, after a few days, she said that the agreement was no longer acceptable. Her rationale: On the basis of the fully clarified facts, she would make different demands.

Joyce made a new proposal, asking for more out of the sale of the house, and more of the antiques. Because she had the same job before, during, and after the marriage, she made no alimony request, her lawyer advising her that no judge would grant it.

As there were no children, there was, of course, no question of child support. Thus, the settlement was restricted to property.

Rick and his lawyer, Jerry, discussed Joyce's new proposal and decided he had three options: stick, give in, or threaten to sue. They figured that Joyce's options were comparable: stick, talk, or threaten to sue. All of the possible outcomes are shown below in Figure 12.

Jerry and Rick must now establish his priorities. Unlike previous examples, however, they will focus not so much on what Rick wants but on where he is vulnerable:

"Rick, the original agreement very slightly favors Joyce. We *could* threaten litigation to try for a better deal.

"The last thing I want to do. I mean, as it is, I feel that I've wasted part of her life. Let's just settle on the original terms."

Figure 12

Joyce sticks	Rick sticks
Joyce talks	Rick sticks
Joyce threatens	Rick sticks
Joyce sticks	Rick (would) give in
Joyce talks	Rick gives in
Joyce threatens	Rick gives in
Joyce sticks	Rick threatens
Joyce talks	Rick threatens
Joyce threatens	Rick threatens

(Guilt feelings, of course, are one of the staples not only of divorce, but of other breakdowns in friendship. Although they may be viewed by some as irrational, they should always be included when establishing priorities. If you leave them out, the assessment will be totally untrue to you.)

"But, apparently she's repudiated the original agreement."

"Well, she might change her mind again."

"Possibly."

"But Jerry, if Joyce keeps nibbling, to hell with her, sue if necessary."

"So, that is what you really want."

"It's not what I want. What I really want is for her to stick to the original deal. Then it's over."

"I see."

"But if she keeps putting her hand out . . ."

"Then suing would be okay?"

"Yes, I want this thing settled as fast as possible, and I don't care how you do it."

"That sounds very grave."

"I mean hire a private detective."

"What for?"

"Jerry, we could prove she's living in my house with another man."

"Are you sure?"

"Positive. Several good friends of ours have told me."

"Rick, this kind of thing can get very ugly. She may have something on you."

"What?"

"Well, perhaps what you've told your psychiatrist."

"That's privileged information—isn't it?"

"Of course, but if you told anyone else about what you told the good doctor, and what he told you, *that* is not privileged."

"I'm in big trouble."

"That's what I thought. Rick, as your lawyer and friend I strongly recommend against threatening a lawsuit or hiring any detectives except as a last resort."

"My point is, if we do litigate, I'll win, even if it's ugly."

"It looks that way. She'd be certain to make a sympathy pitch to the judge, but he probably wouldn't pay much attention. The judges rotate, spending three months a year on divorce court, and usually hating every minute of it. They try to tune out as much muck as possible, just to keep their own sanity. Of course, they have to pay enough attention to get the broad outlines of the case."

"Could we lose if she countersues?"

"I don't think so. At worst, you'd have to pay court costs, winning half of a smaller pie."

"That's still not bad, I'll be out."

"She's being very foolish to have someone move in with her at this point. She's visibly committing adultery —leaving herself without a legal leg to stand on."

"But suing would be emotionally draining, and I'd prefer not to, if possible. I mean, it's basically dirty pool."

"Rick, what if she doesn't threaten to go to court, but just keeps insisting on more?"

"Well, insist on the original deal."

"That could go on for months. As you know, every time a lawyer writes a letter, it costs."

"You don't have to tell me."

"If this thing drags on for long, it could cost about the same as a courtroom showdown."

"But in court I'd have the satisfaction of making a public spectacle of her."

"You shouldn't think that way."

"I don't want to, but I don't want her to turn the negotiation into a Chinese water torture either. I mean, who knows, if she keeps whimpering, I might eventually give in."

"Possibly."

"And, that would be pretty bad."

"Do you really think you've given her enough?"

"Are you insane? Her original demands are too much!"

"I know she's asking for a great deal."

"A great deal for her, not for me. Who's side are you on, anyway?"

"Take it easy. I just want to know if you're willing to . . ."

"Totally out of the question, no matter how she puts it."

"Okay, okay. One last thing, Rick, what if she threatens to sue, but we don't threaten a countersuit? I simply insist on sticking to the original agreement, as a way of quieting things down."

"And let her get away with threatening me?"

"Bad, huh?"

"Why should I put up with that?"

They now have completed Rick's priorities, which we would normally list such that all of the outcomes for each of Rick's strategies are together. However, for reasons which will be made clear later, we shall list Rick's priorities with all of the outcomes for each of Joyce's strategies together, Figure 13.

According to his lawyer, Rick holds the upper hand. However, to enforce his claim may well cost him plenty, not an uncommon situation in areas other than

Figure 13

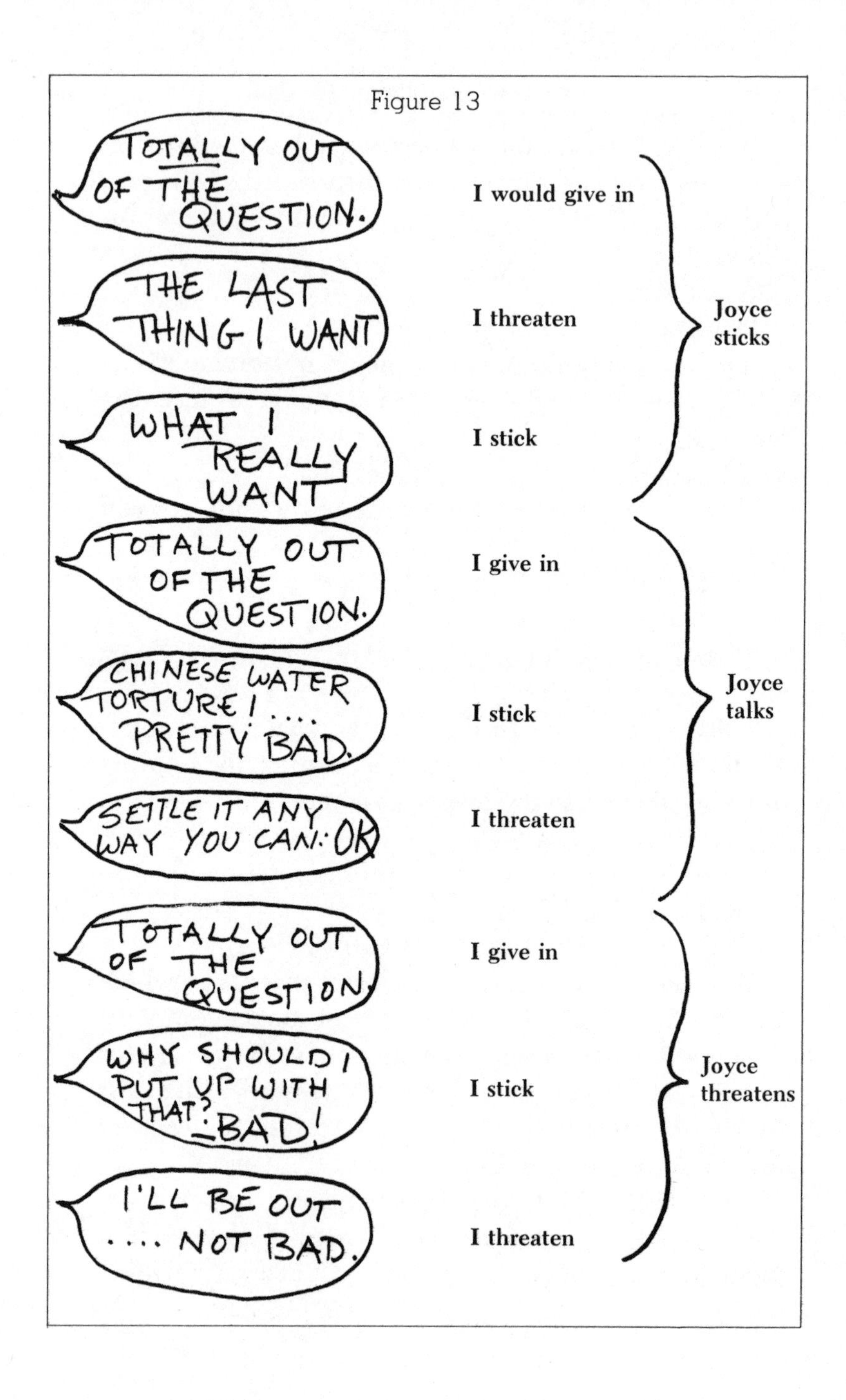

TOTALLY OUT OF THE QUESTION.
I would give in
THE LAST THING I WANT
I threaten
WHAT I REALLY WANT
I stick
Joyce sticks
TOTALLY OUT OF THE QUESTION.
I give in
CHINESE WATER TORTURE! .... PRETTY BAD.
I stick
SETTLE IT ANY WAY YOU CAN.: OK
I threaten
Joyce talks
TOTALLY OUT OF THE QUESTION.
I give in
WHY SHOULD I PUT UP WITH THAT?: BAD!
I stick
I'LL BE OUT .... NOT BAD.
I threaten
Joyce threatens

divorce. For example, a publisher gave a prospective author a huge advance. A year after the deadline for the manuscript the "author" still hadn't delivered. At this point his lawyer got in touch:

"If you sue, it'll cost you ten thousand in court costs, even if you win. And, we can always produce some kind of manuscript before we get to court—if we have to, we'll hire a ghost writer to knock it out. So you might lose the case anyway. Here's what I'm offering: your advance, minus ten thousand, and an additional grand for my fee."

The publisher tells other authors he sued.

In previous chapters, at this point, we have usually worked out the walk-away strategy, because underlying all earlier examples has been a single threat: "I'll leave." But Rick and Joyce have already parted company. The threat to leave is meaningless. There only remains the threat to sue, or perhaps the threat to drag out the proceedings.

In order to fully assess the impact of this kind of threat, a different kind of walk-away strategy will be used, and to work it out we must now look at Joyce's priorities. As in other examples, the other party, in this case Joyce, doesn't establish them herself—Rick and Jerry make some well-educated guesses based on her behavior, their knowledge, and hearsay:

"Jerry, have you heard that Joyce's lawyer tried to be let off the case, but she insisted he stay on?"

"Who told you that?"

"Several mutual friends."

"He objected when she asked for more, telling her she'd be lucky to get what's in the original agreement."

"That's very interesting. Look, Rick, obviously the

best thing for Joyce is to talk you into a new agreement."

"She's living in never-never land."

"Maybe, but that's clearly what she wants."

"Sure."

"And I think she would see it as pretty good to drag out the negotiations ad nauseam, figuring that she'll eventually wear you down."

"She has outwaited me once already—when I moved out."

"And she might figure you'd give in at least a little if she threatens."

"But, it's an absurd possibility. I wouldn't do it."

"She'd like you to. She'd feel okay about it."

"Yes."

"Of course, if we countersue, she'll lose big."

"Well, that's obviously the worst for her."

"I think so, unless she'd enjoy making you miserable in court."

"She might enjoy it, but she's not so crazy that she prefers sadism to money, I'm sure of that."

"It does happen, Rick."

"If she sues or even threatens, we can beat her in court without a countersuit; we just propose her original agreement as the settlement—a straight division of community property, and her signature's on it."

"You never can tell, it would be up to the judge to sort it out, which is why this isn't bad for her. Who knows? She might get something more, and at least she'd feel she left no stone unturned."

"But it's Joyce's own proposal!"

"Of course. But regardless of what she does, if you

do sue her at this point, she's finished. Since she's living with someone in your house, it utterly demolishes her case as the wronged woman. By the way, how do you feel about her and this guy?"

"Oh, I'm glad she's found someone—it lets me off the hook. Maybe she'll be happy; I hope so."

"Rick, she's still seeing an analyst, isn't she?"

"Yes, and I'm still paying for it."

"And you should keep paying until this thing is settled. If we have to litigate, it'll be to our advantage."

"I suppose so."

"Do you suppose her lawyer has discussed alimony with her?"

"I won't pay alimony! Not one penny. I'd leave the state first."

"Not alimony for her, for you."

"For me?"

"In this state, alimony sometimes is awarded to the husband if the wife has had a steady income and the husband has not."

"Are you kidding?"

"No."

"Well, I don't want any alimony."

"Good, but her lawyer has probably explained the danger of it in any litigation. Rick, do you know what's driving her to make these new demands?"

"What?

"Somebody, I don't know who, keeps telling her, 'You'll regret settling now; Rick will give you more if you just keep at him. Maybe he feels bad about things, feels guilty. He certainly should! You supported him all those years.' "

"Sure, she supported me for a couple of years. But what about the other years, when we traveled first class all over Europe. All but one summer we had a suite on the *France.* Could she have done that on her third-grade teacher's salary?"

"Regardless. My point is the fear that she might be passing up a golden opportunity is what's driving her on."

"Well, I can't blame her. Missing out on something she feels—however erroneously—is hers would obviously be bad."

"Which is why settling for the original agreement is also pretty bad."

Rick and Jerry have now established what they believe are Joyce's priorities. They list Joyce's priorities so that all of the outcomes for each of *Rick's* strategies are together, see Figure 14.

What should Rick do, based on this analysis? Suppose Rick and Joyce not only stop talking altogether, but sue each other. Joyce gets the "worst" outcome while Rick gets one which is "not bad." Once they get to this outcome, they're stuck with it, because neither has any incentive to change. Joyce is no better off by dropping her lawsuit if Rick keeps up his, since she would continue to get the "worst." Rick wouldn't change if Joyce doesn't, because his value goes down from "not bad" to either "bad" or "totally out of the question." So they would be locked in.

The only other outcome which is as rock stable is if Rick threatens and Joyce talks, again they would be deadlocked. Any change would cause Rick to go from an outcome which is "okay" to one which is either

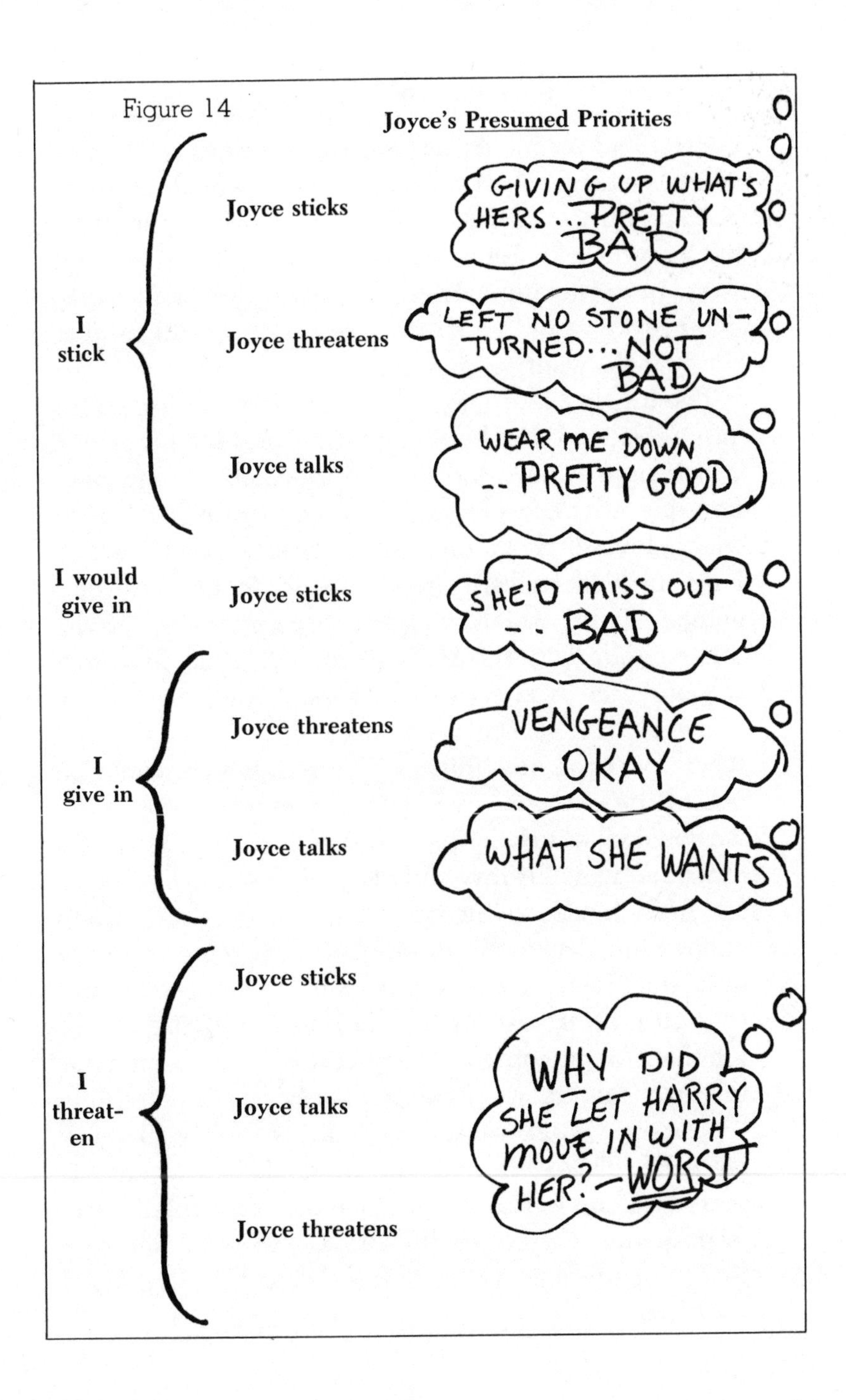
Figure 14
Joyce's Presumed Priorities
I stick
Joyce sticks
GIVING UP WHAT'S HERS... PRETTY BAD
Joyce threatens
LEFT NO STONE UN-TURNED... NOT BAD
Joyce talks
WEAR ME DOWN -- PRETTY GOOD
I would give in
Joyce sticks
SHE'D MISS OUT -- BAD
I give in
Joyce threatens
VENGEANCE -- OKAY
Joyce talks
WHAT SHE WANTS
I threat-en
Joyce sticks
Joyce talks
Joyce threatens
WHY DID SHE LET HARRY MOVE IN WITH HER?—WORST

"pretty bad" or the "totally out of the question." If Rick doesn't change his strategy, Joyce cannot gain by changing hers. Notice, however, that if Rick threatens but Joyce sticks, Rick would now want to change his strategy to sticking with the original agreement. However, Joyce would now rather switch to talking, and so Rick would want to change.

The lists in Figures 13 and 14 have been arranged for easy selection of stable outcomes. Consider Figure 14, which shows Joyce's presumed priorities. We see all of her different values for each of Rick's strategies. If Rick sticks, Joyce's best outcome is "pretty good," which happens if she talks. If Rick gives in, Joyce's best outcome is "best," which again happens if she talks. On the other hand, if Rick threatens, she could do anything, the outcome is always the "worst." Naturally, if she knew what he was going to do—and it's no secret—she would pick the best thing for herself, which would be one of the outcomes we have just enumerated. Then, we could look at Rick's priorities, Figure 13, to see if this outcome, which is favorable to Joyce, would also be so for Rick. Since, given her choice he might want to change his. For example, suppose Joyce would like the outcome "Joyce talks/Rick gives in," which, according to Figure 14, is best for her. Consulting Figure 13, we see that this outcome is "worst" for Rick, who, on hearing Joyce talking, would immediately start threatening.

Suppose we use mutual lawsuits as their walk-away outcome. This is, in fact, the most likely and natural such outcome in the whole situation, the kind of nasty dispute most frequently associated with divorce. Joyce's walk-away value is the "worst" and Rick's is "not bad."

As in other examples, any *negotiated* outcome would have to be better for both of them. This immediately eliminates all lawsuit outcomes—Joyce would only get her walk-away value. Also eliminated would be any concession on Rick's part, since this gives him an outcome worse than his walk-away value. The outcome where Rick stands his ground but Joyce threatens suit is out for the same reason, as is the one where Joyce drags out the bargaining. In fact, all that's left is for both to settle for the original deal! So, Rick holds the high cards.

There is a subtler game-theory solution (that of Nash) which perhaps should be applied to this example. But, this alternative solution cannot be explained nonmathematically. So it won't be explained here. However, if we were to apply the other procedure, we would find that Rick should expect a certain amount of delay while Joyce keeps asking for more.*

The solution we have examined shows who should make the concession—Joyce. This fact was based on the relative damage they could do to each other if they both sued.

Cooperation is possible in the absence of an overriding common interest. Both parties must be made aware of how destructive they can be to each other, and to themselves in trying to do the damage. The critical contribution of game theory is to make sure that one is aware of his or her own weaknesses, as well as those of the other party. Then you probably won't rush into foolish threats, which might later be regretted.

The fact that there is more than one possible game-

*Assuming their priorities are represented by consecutive integers.

theory solution is critical. Sometimes, no amount of logic will show the One Way to resolve a dispute—because there are more ways than one! Forget about *the* solution, and instead think about behaving reasonably, and try not to squeeze the last drop of advantage from the situation. The main thing is to do better than your walk-away value.

Rick made one mistake. He should have tried to communicate the critical strategic facts to Joyce; of course, she might not have wanted to listen.

How did things end up? Neither side litigated and they eventually settled for the original agreement. But, the additional legal fees of a protracted negotiation—one and a half years—ate up some of the value of the community property. Did Joyce feel she was worse off than she would have been with an early settlement? Obviously yes! Did Rick? Of course. The final result was permanent anger; they haven't spoken to each other for over two years. Thus, a bitterness developed which was not present at the time of the actual separation, and was not provoked by the actions of either lawyer.

Had Joyce analyzed her situation at the beginning, she might have faced up to the futility of dragging out the proceedings.

* * *

This chapter has shown how to deal with someone who has repudiated an agreement.

OPTIONS

First, determine your options, and those of the other person.

YOUR VULNERABILITIES

Second, establish your priorities. Pay particular attention to where you are vulnerable, not only to the other person, but in terms of your own regrets and guilt.

HIS OR HER VULNERABILITIES

Third, try to figure out the other person's priorities, again focusing on vulnerability.

HIS OR HER STRATEGIES

Fourth, make sure that your priorities are listed as in Figure 13, that is, by the other person's strategies. Similarly, make sure that the other person's priorities are listed as in Figure 14, that is, by your strategies.

DAMAGE

Fifth, find the stable outcomes, as explained in the text. One of these may be an obvious "walk-away" outcome. If so, this outcome represents the maximum damage that both parties can do to each other, and is what you both get if you don't come to terms.

BEST DEAL

Sixth, using the outcome determined in step five as the walk-away value, work out the best deal, as explained in Chapter 6.

TALK

Seventh, make sure the other party is informed of the advantages of a cooperative solution as opposed to a nasty one.

9

Make up His Mind

Union leaders sometimes use their inability to control the rank and file as a bargaining threat. At a critical moment in the negotiation, after management has made an offer which is almost satisfactory, the union "boss" says: "We'd like to go along with your offer, but we're just not sure if the men will buy it. If they don't you'll have the worst strike imaginable—a spontaneous rank-and-file movement. You know the trouble that would be to halt, and you can guess what the settlement would have to be. So why not sweeten the terms now just to make sure."

A similar kind of threat can appear when two partners plan to work together on a business venture. At last the bank loan is available, but one partner suddenly isn't sure if he wants to go ahead. He doesn't say, "No deal. I've changed my mind." He just stalls for time.

Now his wishes dominate the considerations.

Trivial instances of this occur everyday. For example, two people plan to see a movie, but at the last minute one isn't sure: "I've heard this picture stinks." Of course, the classic examples are stock Hollywood: The bridegroom dragged to the altar by the best man, or the bride who locks herself in the bathroom seconds before the ceremony and refuses to come out.

Did the one with cold feet think from the beginning, perhaps subconsciously, that the event wouldn't really happen? Intrigued by the dream, the falterer is scared by reality: "Wait a minute, do I really want to go through with this?"

This chapter will show one effective way to answer the threat, whether deliberate or not, of a hesitating partner.

* * *

George and Dick have spent the past year working on a book proposal, into which they have put a great deal of time, energy, and research.

The collaboration seemed perfect—one had knowledge the other lacked, and vice versa. Dick is an expert in the field but has never written a book. George, on the other hand, has no knowledge of the subject, but is a semiprofessional writer, having written several previous books in a different and highly specialized area. They are also good friends who work well together, without rivalry or jealousy. Since the original idea came up in conversation, and neither remembers who first suggested it, they feel the proposal belongs equally to both of them.

They get an agent, Ellen, who warns them not to inflate their hopes for an unrealistically big advance: "An unwritten book from unknown authors; it's a matter of luck." Ellen sends the proposal around to a number of publishers, one of which shows some interest. George and Dick meet with the editor from the publishing house, who subsequently makes an offer.

The offered advance is several thousand dollars below the figure to which they had jacked up their expectations. Ellen advises grabbing it: "They're a large and prestigious publisher," an important consideration to Dick, who needs academic respectability, but of no significance to George. The book is out of his field.

The contract is drawn up and sent to them. Dick is enthusiastic, having not only reconciled himself to the small advance, but celebrating the arrival of the contract with a dinner party. Not so with George: "Dick, you've never written a book before; you don't know how much work it involves. I've written three. Each required several nitpicking revisions. Finally you give it to the publisher, who doesn't have to accept it. They can always say, 'This isn't the book we had in mind.' After all, they've only seen a proposal. And then, maybe no other publisher will want it either. We could literally be stuck with a finished book. All we'd get out of it is that piddling advance, which we have to divide in half, and out of which our agent takes ten percent. Peanuts for a year's work, and all of our extra time—two, three, even four evenings a week together, plus the outside research. Twice as much money might be reasonable pay. But with this advance, I'm just not sure."

"If you want reasonable pay, get a job working for the

city. With this book we could get rich."

"Dick, we don't have to sign this minute. Let's sit on it for a few days to give me some time to think it over. Let's read the contract through carefully to make sure it has everything in it that's supposed to be there, and nothing that isn't."

Dick gets angry. When he started the project, he assumed George would see it through to the end. Now, Dick feels let down. Furthermore, he's sure the publisher would be scared off by the slightest hint of hesitation. Collaborative projects often get nowhere, even after the contract is signed; the books are never finished because the authors end up hating each other. People have already joked: "Are you and your coauthor still speaking to each other?"

Thus, because George is acting indecisively, Dick is also thrown into turmoil, feeling that George has been frivolous about the whole thing.

The options of Dick and George are perfectly obvious and straightforward—each can sign or not. The four possible outcomes are shown in Figure 15.

Dick establishes his priorities in a conversation with his wife, Vera: "I want to get started as a writer, so having a publisher now is a real break. But George thinks if this publisher's interested others will be, too, and if we wait we might get a better offer."

"Do you think he's right?"

"I'm not so sure."

"Would the publisher accept you without George?"

"Maybe not, but with a few sample chapters, I might convince them."

"Yes."

Figure 15

George signs	**Dick signs**
George signs	**Dick rejects**
George rejects	**Dick signs**
George rejects	**Dick rejects**

"In any case, if George is hesitant, imagine the trouble even if he does sign: 'I can't work today; don't bother me now; call me next week.' The project could drag on for years. So, I have my doubts whether or not I want to work with George at all."

"You do?"

"Yes. I didn't realize it until now, but taking the offer is more important than continuing the partnership. Anyway, suppose in a few months another publisher offers twice as much; George might still say it's not enough.

"So you feel it's best to do the book on your own?"

"Not really. In spite of what I've just said, I'd still rather write the book with George—providing we write it now."

"So it's best if you both sign, but doing it alone is okay?"

"Right."

"And going along with what George wants . . ."

"Is bad."

"What if he finally signs but then doesn't do any more work?"

"I don't care, I want the contract. I'll write the book by myself if necessary—the publisher won't know until it's finished, if then."

"You don't mind giving him half the money for nothing?"

"Of course I do, but it would be worth it because then I'd have the contract. Anyway, I'd get some work out of him. He's not a crook. He just doesn't know what he wants to do."

"Do you think he's trying to shake you loose so he can write the book alone? If he accepted this offer by himself, he'd have the full advance, twice as much as he's now offered."

"Yes, he would, wouldn't he? But I don't think he wants to write it on his own, and even if he did, I can't believe he'd pull something like that—although people who have worked with him in the past have warned me about him. You never think it'll happen to you, but sometimes it does."

"So this would be the worst?"

"You said it!"

Dick has now established his priorities, see Figure 16.

Notice that Dick includes a paranoid fear—a double-cross by George. Some would say that including this is irrational, and Dick should look at the problem strictly in a dispassionate way. But Dick isn't dispassionate, and game theory tells him to include his passions in his priorities. This is very important. Dick must have no uneasiness about the emotional validity of his walk-away value when he talks to George.

THE WORST?
YOU SAID IT!
George signs / I reject
George rejects / I reject
BAD.
OK?
RIGHT.
George rejects / I sign
George signs / I sign
BEST!

Figure 16

Always include in your priorities whatever is on your mind, regardless of how crazy other people say it is. You may forget the logical reason for what you do, but your feelings will linger, whether you want them to or not.

George's priorities are quite different, as revealed by his conversation with his friend, Irving:

"Irving, a year ago, I was very enthusiastic about this book. And at that time I might have been happy with a small advance from a good publisher. But things have changed. I have a lot more meat on the plate now."

"The book doesn't interest you anymore?"

"Sure it does, but not like it once did. Of course, if the advance were twice as big, that's a different story. And I think we can get it if we sit tight."

"That's what you really want?"

"Sure, send this proposal on to more publishers and see what happens. We've had a definite offer from a good publisher; this gives us a little bargaining leverage. Maybe Ellen can auction the thing. Dick's running scared: he should be willing to gamble."

"Let me get this straight, George. Would you mind if Dick picked up the ball and ran with it alone?"

"That's the worst thing that could happen. This book, after all, has the potential to make bags of money. How many chances like that have I had? Also, it's an interesting project, something that has intrigued me for years."

"You surprise me, George; I was certain you'd say, 'Surely he wouldn't think of doing it without *me.*'"

"That, too. Look, the point is I don't want to be the one left out in the cold. If we're going to break up the team, I'd rather have the contract and let him go his own way."

"You want to do that?"

"I didn't say I wanted to do it, what I said was, I'd rather he missed the bus than I did."

"Well, that's fair enough."

"Yes, but I don't want it to happen; it's still bad."

"Look, George, it sounds to me like going along with Dick is okay."

"Looks that way, doesn't it? Irving, I don't like settling for second best."

"What are you going to do?"

"Some persuasive talking."

"A con job? It won't work."

"Not a con, Irving. Everyone does what he ultimately wants to do."

George's priorities are shown in Figure 17.

Dick's priorities, Figure 16, show a clear walk-away strategy. The best outcome is for both to sign. But, second best—which is okay—is if Dick signs and George doesn't. Both outcomes are preferred to rejecting the offer, regardless of what George does. The worst that can happen if Dick signs is okay but the worst that can happen if he rejects is the worst possible. Thus, "okay" is Dick's walk-away value, and signing is his walk-away strategy.

Of course, he would like to persuade, or at least prevail upon George to go along, but Dick's walk-away value is quite high, the second-best outcome. If unable to get George to sign, Dick should proceed by himself and take his chances with the publisher.

Ironicially, George's walk-away strategy is also to sign, but he's not happy about his walk-away value. As shown in Figure 17, the worst that can happen if he

Figure 17
THE WORST THING.
I reject/Dick signs
WHAT I REALLY WANT!
I reject/Dick rejects
IT'S BAD
I sign/Dick rejects
OK
I sign/Dick signs

signs is "bad." But, the worst that can happen if he rejects is the worst possible. Naturally, he'd rather have a bad outcome than the worst one, so as a last resort, he should sign. However, the fact that his walk-away strategy is to sign undermines the case he plans to make to Dick that they should keep sending the proposal around.

This ambivalence holds him completely tongue-tied, despite his confident boasts to Irving. Finally, Dick forces a meeting to resolve the matter, which has to be settled one way or the other.

"Look George, this publisher means a lot to me. I know the advance isn't as good as it should have been. They never are. It's good enough for me, and I'm going for this contract—by myself, if necessary. I wish you'd come in on it, but I'm not going to waste time waiting for a better offer. Of course, the publisher may not want me by myself. I don't know. That would be too bad, but I'm not going to lose it through any fault of my own."

George, at first at a loss for words, quickly recovers: "Okay. It's not what I really want; I'd prefer we turned it down flat. But I certainly don't want to leave you in the lurch. So, I'll sign, Dick, because I know this means a lot to you."

Dick hands him a pen.

"But, there's one proviso, Dick."

"What?"

"I set the work schedule. I don't want you pestering me every twenty minutes to work on this book."

* * *

George and Dick couldn't both get what they most wanted. Dick, in fact, gets what he wants while George settles for second best. However, George feels he comes through with a built-in alibi. If the book takes off, he can point to his wisdom in signing. If it flops, he can blame Dick: "We should have waited for a bigger advance, at least we'd have *that* money."

Dick didn't call George on his claim of selflessness because George may have believed his own rhetoric, so why offend him? Furthermore, Dick is more than happy to accept any future blame for failure; it amounts to nothing, anyway. Finally, since George "controls" the schedule, he feels he gets something from the bargain, but this is such a vague concession that Dick feels he can subsequently wheedle his way around it.

The main point: George's uncertainty was answered by Dick's decisiveness. He made up George's mind. If George hadn't already clarified his own priorities, however, the ploy might not have worked. The delay in signing turned out, to the surprise of both, to benefit both.

Many suddenly indecisive people are in George's position: torn between preserving a relationship and rejecting an offer. If you answer them with decisiveness, they almost have to follow your lead—or they make themselves even more miserable.

If you knuckle under, you may make them happy, but yourself miserable.

In effect, Dick is using a punishing strategy, since it forces George into line, unless he is a glutton for punishment. However, Dick's strategy is not manipulative because it is also a walk-away strategy. Dick is simply living his own life. If George wants to go along for the ride, fine.

* * *

This chapter has shown one way to deal with a suddenly indecisive partner.

DELAY

First, each person must determine his or her priorities and walk-away strategy. This requires giving the indecisive partner time to figure out where he or she stands. Although this may seem like a delay, the time will be well spent in the long run.

ON YOUR OWN

Second, if feasible make clear that you will proceed with the project on your own. If you can't believably make this claim, restate the issue as one in which you are determined to do what you want to do among the available options. If your partner's priorities are such that he or she would rather go along than go away, you've settled the matter. If his or her priorities are not like this, you can forget about the partnership anyway.

SMALL CONCESSIONS

Third, consistent with what you want to do, make minor concessions, be polite and make your partner feel that his or her wishes haven't been completely steam-rollered.

10

The Poor Man's Headhunter

The go-between is a classic figure in literature. Recently, a modern form of go-between, the headhunter, has been widely discussed. This chapter shows when you need a go-between, and how to size up a potential one.

Alcibiades, the Athenian general and military genius during the Peloponnesian War of ancient Greece, knew the secret of changing jobs. He became disenchanted with the Athenians and went over to the enemy, the Spartans. They welcomed him not only for his Athenian military secrets, but for his talent. The war dragged on. Alcibiades, now older and wiser, and disenchanted with the Spartans, went back to the Athenians. And they took him back. He was talented, a turncoat with a silver lining.

In our own time Wernher von Braun has had an

analogous career. He developed the Nazi V-2 rockets which rained destruction on London in the closing days of World War II. Shortly before the final Nazi defeat, he surrendered to the Americans, and was instantly hired by the U.S. Army for its missile program.

Years later, von Braun was glorified in the movie, *I Aim at the Stars.* Not everyone was willing to let bygones be bygones and forget his Nazi past. As the London *Daily Mirror* put it, "He aimed at the stars, but he hit London."

There is no question but that von Braun made a smart career move. The Russians probably would have put him to work, but he had strong doubts about living conditions, pay, and, in particular, duration of contract. The British, on the other hand, might have kept him cooling his heels for years, never offering him a satisfying job.

The refusal to hire someone solely because he or she works for or has just quit a rival firm, is called an "antiraiding" policy. Although no bar to recruitment for the U.S. missile program, such policies sometimes block mobility in more down-to-earth fields, such as art.

Before he became world famous, Al showed at a small, but very reputable gallery. However, he was driven by ambition, and so wasn't satisfied, wanting to advance to a more prestigious show place. The obstacle was the owner of the desired gallery, Lee, who wanted to show Al's work but wouldn't, under any circumstances, violate the unwritten injunction against raiding. Lee had a well-deserved reputation for being totally honest. But he was also no dummy, and wasn't about to let this artist get away. Something had to be done fast because a less scrupulous dealer, not bound by a sense of fair play, was already sniffing around.

The original dealer, Jim, simply couldn't afford to advance the money the artist wanted. Jim, Lee, and Al were friendly, often dining together, at which time Lee played on Al ambitions: "You should have everything money can buy."

The upshot: Lee and Jim both paid a monthly advance to Al, who, to this day, shows at both galleries. Lee got what he wanted without raiding. Jim, the potential raiding victim, maintained exactly what he could afford, and so retained a valuable piece of the action. And Al advanced his career, but kept an old friend at the same time.

The artist was lucky because the business world in which he operated—the art world—is small and personal. Everybody knows everybody. On the one hand this constrains particularly nasty behavior, because one can easily be branded; on the other hand, the chumminess allows for flexibility.

Much of business, of course, is more cut-throat—even if ostensibly restrained by antiraiding policies. Potential raiders don't want to admit their conduct unless it succeeds, and ambitious job changers don't want to make any obvious boo-boos either—their careers could be killed or stunted. The situation is very close to that faced by Alcibiades. The solution? A totally confidential go-between.

At the top levels of management and some professions, such as law, the hush-hush middleman is called a headhunter. He or she is an employment agent who finds or sizes up prospective executives for new jobs before they even know they're being considered. How can this help the guy who wants to change jobs?

He should cultivate the friendship of a headhunter, doing favors for him, suggesting the names of particularly talented people, and informing him of companies that might be in need of his services. In exchange, the headhunter keeps this middle-or senior-level executive in mind as a possible candidate for a job change. Of course, the latter is never discussed, and there is no hint of an actual tit for tat. But both know what's going on.

Headhunters charge a lot of money and at least claim to operate with the flair, if not the violence, of a James Bond. So executives enjoy being scalped.

But changing jobs isn't much fun for someone way down the ladder, where there isn't enough money in the commission to make the hunt worth outfitting. Who is the poor man's headhunter? A friend, of course. The problem is finding a suitable one, or sizing up those who are already available.

Cultivating friendships, although an interesting topic, is not the focus of this book, but sizing up a possible go-between is clearly a game-theory problem.

* * *

Joe, a young lawyer working for *the* major Wall Street tax law firm, is unhappy with his job. Although he earns a fair salary, he simply doesn't get the juicy cases he feels he needs to advance his career. He does extremely well at what he gets, but there are other young lawyers in the firm who also do good work; the best cases are divvied out among them. Furthermore, whether true or not, he feels that once he solves a problem, the senior partners never know who did it because the credit is ripped off by his immediate boss. As there seems to be no practical way to short-circuit the information flow, he feels his chances for rapid promotion to the ranks of partnership are very poor.

Joe's had his eye on a smaller firm where he feels the action would be faster: "Given a chance, I'd be their number one young lawyer." He feels that the smaller firm would offer him a greater opportunity for advancement, since he'd be recognized for his work.

Joe's present firm deals exclusively with corporations, whereas the smaller one handles celebrities. A job at the other firm would involve a direct contact with the clients, which Joe does not have now, and as a bonus, the clients themselves would be glamorous and exciting.

Of course, he doesn't even know if there's an opening, but he figures if they want him, they can make one. In the course of his work, he has been in contact with one of the senior partners in the smaller firm. But Joe doesn't know the best way to follow up. He does mention to a number of friends that he thinks getting ahead in his present job will take a long time: "Everything takes time. Of course, I wouldn't mind speeding it up a bit." So, he's let the word out generally that he's eager.

The question is: How should Joe approach the firm? He doesn't want to appear over-eager, but he doesn't want to waste his life either. He eventually decides that he has three options: do nothing; make a direct personal contact; or ask a friend, Hal, who plays bridge with the partner, to act as intermediary. Although his third option seems appealing, it has to be carefully weighed against the alternatives; the partner may play bridge with Hal, but also think he's a dope.

The other firm has various options. First, they can make clear that they're not interested: "You? Work for us?" Second, they can advise him to leave his present job, but make no offer: "You're not happy? Quit!" Third, they can open discussions, perhaps with a meager offer: "You're worth more? Prove it!" Fourth, they can make a generous offer: "Son, here's the key to the executive washroom."

Since there is no way for Joe really to assess, in advance, the other firm's reaction to his making contact, he should proceed with caution, which means using a walk-away strategy.

He does not have to evaluate the entire list of outcomes in Figure 18, but, instead, only has to look at the worst outcome for each of his choices. He needs the entire list, however, to inspect for jokers in the deck.

Joe talks the situation over with his wife, Alice: "No question, the worst thing would be to go hat in hand, and they advise me to quit my present job, period."

"And they leave it at that?"

"Yes, they don't offer me anything."

"They'd have to be pretty perverse if they didn't subsequently offer you *something.*"

"Yes, but what and when?"

"Well, they could be saying 'Make yourself absolutely available, and *maybe* you'll hear from us,' or else they might simply mean 'If you're unhappy, put yourself on the market.' "

"Right, I wouldn't know if they're giving advice or hinting at an offer."

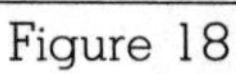
Figure 18

I do nothing

We're sure glad you're not working for us.

You're not happy? Quit!

This is all we offer. You're worth more? Prove it!

Son, here's the key to the executive washroom.

I make direct contact

You? Work for us?

You're not happy? Quit!

This is all we offer. You're worth more? Prove it!

Son, here's the key to the executive washroom.

I send Hal

Joe? Work for us?

He's not happy? Let him quit!

This is it. He's worth more? Let him prove it.

Tell him his key is waiting.

"They could also be saying, 'We aren't raiders. If you were a free agent, we'd hire you.' "

"But, I could end up just being a free agent."

"True."

"Even if I didn't take the risk, the whole thing could get back to my boss."

"Yeah, you don't want him to find out. What if in the course of your regular business, the other firm's senior partner advises you to stay where you are?"

"Probably wouldn't happen, but if it did, it would be pretty bad, too. My grumbling would have become common knowledge."

"Yes. On the other hand, *they* would be taking a paternal interest, and *you* wouldn't be seeking another job. They'd merely have heard via the grapevine that you're restless."

"Which means the grapevine is broadcasting loud and clear."

"At least they're keeping an eye on you, which is always good."

"Mainly, I know I shouldn't do anything, so it clarifies that."

"Yes, and it's a lot better than being caught out on a limb."

"Suppose Hal gets in touch, and they say 'No, thank you.' "

"Maybe they would be advising you to stay where you are because there would be faster advancement than you think."

"I doubt it."

"They may know something you don't about what's happening in your firm. Sometimes the top guys in other firms do."

"Regardless. The point is if anything goes wrong, I can always divorce myself from Hal."

"Are you afraid he can't handle it?"

"Well, we can work it out if we talk it over carefully first. I mean, for all I know the partner doesn't respect Hal. But I know they're friendly enough to play bridge together. So Hal should put it very casually, like: 'I think my friend Joe is restless, and I thought you might be interested in knowing.' That would be a lot different from saying 'I know this fantastic hotshot. You guys ought to grab him!' "

"You're right, it should be done in a way that can't backfire."

"It's the best option, but still not good unless they offer something."

The key, as far as Joe is concerned, is whether or not Hal can handle the matter. Any intermediary is not necessarily better than no intermediary. However, Joe has decided that Hal can deal discreetly with the contact, providing they talk it over first. Sending Hal thus becomes Joe's walk-away strategy. But, had Hal not been up to the task, Joe's walk-away strategy might have been to do nothing.

Joe and Hal meet to discuss the firm's possible reactions: "What if they advise me to quit but don't make an offer?"

"It's sometimes done, but the danger is that things could come up suddenly that they couldn't foresee, such as a sudden loss of major clients."

"Yeah, I hadn't thought of that."

"In any case, they'd certainly get in touch with you before you did anything. You could feel them out then."

"Suppose they hint at a crummy offer?"

"It would depend on the offer. A meager salary in a lousy position is one thing, but a meager salary with possibilities for rapid advancement is something else. A small offer might be their way of opening negotiations. Or, it may be all that's going. I'd try to find out."

"First offers are rarely final, even when people say they are."

"Right, but maybe they'll make a generous offer. Then they'll simply want you—without any bullshit. Of course, they might just ask for you to get in touch with them, in which case, it's your baby."

"But if they advise that I stay where I am, that would be it."

"Maybe not. They're not quite as sure of themselves if they operate through a go-between."

"Yeah, but at that point, it's not worth any further risk on my part."

"It could still mean they might consider you in the future."

Joe and Hal made no attempt to work out the firm's priorities, which would have been impossible. The most they could do was to try to make sense of the firm's possible replies.

Joe got the job.

* * *

This chapter has shown why headhunters are in business. However, they are not always available, and one may consider the use of a friend as an intermediary:

FIND THE ONE

First, find a possible go-between.

YOUR OPTIONS

Second, consider other options, including doing nothing and making a direct contact.

FIRM'S OPTIONS

Third, consider the widest range of options of the firm that interests you.

THE LIST

Fourth, make up the usual list of options.

DISOWN

Fifth, establish your walk-away strategy. You only need to look at the worst outcome for each option. The others are on the list to make sure you don't get taken off guard. You may decide that the prospective intermediary is worse than no intermediary. He or she should have the right social distance from you so that you can disown the contact if anything goes wrong.

TALK

Sixth, if you decide to use an intermediary, be sure to talk things over thoroughly before the contact is made.

11

I Give You My Word

Were you surprised in July of 1976 when ultraconservative Ronald Reagan announced that if the Republicans nominated him for president, he would pick ultraliberal Senator Richard Schweiker as his vice-presidential running mate? You might have been. Both men had gone back on their word. Reagan reneged on his assurance of an ideologically pure ticket. Schweiker, until the morning of the announcement, had pledged his vote at the Republican National Convention to President Ford.

Reagan could have taken Schweiker's turnabout as evidence that his word was suspect, and vice versa. But rightly or wrongly they felt they had an overriding common interest. Reagan hoped to clinch the nomination with the votes of the Pennsylvania delegation. Schweiker, a senator from the state, was expected to deliver the votes.

The senator turned against Ford because Reagan offered a better deal. But that's just politics.

Surely, in nobler areas of human activity, such as love, things are different. What makes a marriage promise believable? A shotgun? Maybe. But assume the couple is in love.

Ask the bride on the way to the altar how she knows the groom won't stand her up. She probably won't say, "Because he gave me his word." She'll say, "Because he loves me." Love is what makes the promise believable. He's made clear to her that he'll be happier married than not married, and happier with her than with anyone else. Presumably, she's made similar feelings clear to him.

So, whether the promise is a political offer or a marriage proposal, the key to its believability isn't trust, but that there must be no better or happier arrangement in sight.

This sounds like an auction, which does work the same way. For example, Harry is selling his old Buick, but will keep it unless he gets at least $1,000. Of course he doesn't tell this to potential buyers. If they know his reserve price, they merely have to offer it.

Suppose the first bidder is prepared to pay up to $1,500, while the second will go as high as $1,800. Naturally, they keep these figures to themselves, hoping to pay less. The first bidder makes his offer: $1,100. The second one replies with $1,200. They keep up the bidding until the price goes above $1,500, at which point the first bidder, no longer capable of making greater offers, goes home. Any amount greater than $1,500, but less than $1,800 is a potential sale, and is in the *core.* This is composed of all possible deals which no group, subgroup, or individual can spoil with a better offer. In Harry's case the core is filled with all possible offers above the cutoff level of the first bidder but below that of the second. Since there are no other bidders, no one can spoil the potential sale with a better offer.

The second buyer and Harry will presumably come to terms, unless the auction has so incited Harry's greed that he now hopes for something more than $1,800. Perhaps "$2,000" suddenly lights up in his eyes. The second buyer can't pay it, so he leaves in disgust. This nice round sum isn't in the core.

Successful political deals, happy marriages, and honest used-car auctions have more in common than one might think—all are in the core, meaning everybody's satisfied that they couldn't do better.

This is true even when the promise involves giving something away. For example, an artist made a gigantic piece of sculpture, suitable only for exhibition in a large museum. A millionaire interested in both the art and a big tax writeoff offered to buy the piece at a good price providing a prestigious museum would be willing to accept it. Although the artist was famous, and virtually any museum would have been happy to have a piece of his sculpture, the colossal size of the proposed gift made it a different story, especially since the donor insisted on permanent display. He liked the piece, but he also liked to walk through a museum with friends and casually mention, "Oh yes, I gave them this."

Modern art museums like to periodically change their displays, so when the piece was offered to a major one, its directors hesitated at the commitment. Although the artist was not really a participant in the deal, having already accepted the terms of the conditional sale, he anxiously sweated out the delay while the museum dallied. The donor's wife, meanwhile, got impatient.

Without notifying the first museum, she offered the piece to another one she hoped would be more eager. When the artist secretly informed the first museum of this fact, they quickly grabbed the piece for fear not only of losing it, but of losing future gifts from the donor as well.

As things turned out, the second museum didn't have room to keep the piece on permanent display anyway. The agreement between the donor and the first museum was clearly in the core—the second couldn't meet the terms of the deal and the first figured it better do so before some other museum offered to put the donor's name up in neon lights; the first museum was in the core only so long as there were no hungrier takers.

The core thus can be used as a test to see if a promise is believable. If in the core, count on the promise, if not, watch out. This kind of question is often easier to figure than whether or not someone is trustworthy.

Besides, if you look at things this way, nobody's ever a liar. People merely say things occasionally which are not in their best interests.

NOTE: *How to Conspire to Rig an Auction*

If the two bidders on Harry's Buick got together, they could hold down the bids below the first one's cutoff price. Why below? Above this figure, the second bidder doesn't need to make a deal. Any amount between the first bidder's cutoff and Harry's reserve price could be a possible settlement. Then the two buyers could split the difference between the actual sale price and the maximum amount the second buyer is willing to pay, $1,800. Why not the maximum the *first* bidder is willing to pay? The first bidder can always hold up the second to this level, so the payola has to be figured on what the second bidder is saving. Of course, he may suddenly come up with a much lower cutoff level than was previously thought.

The result is the first bidder gets a bribe, and the second gets the car. To illustrate, suppose they settle for $1,200 which Harry gets. The first bidder gets a bribe of $300 and the second bidder gets the Buick at $300 less than the maximum he was willing to pay. Of course, by rigging the auction, they're not in the core—because they're not making honest bids.

* * *

This chapter has shown a way to test the reliability of a promise.

BEST DEAL?
First, look at the best possible deal that the promiser could make.

WITH YOU?
Second, see if this best possible deal is with you.

SPOILERS?
Third, if it's not, watch out. The promise isn't in the *core* —all deals which nobody can spoil with a better offer.

GRAB IT

Fourth, the deal may be in the core, but not for long—somebody else might offer a better one. Grab it fast.

MOTIVE

Fifth, remember the promise may be true even if it's not in the core. But, if in the core, the promiser has a clear motive to tell the truth.

12

Moving On

The previous chapter gave a yardstick by which to measure the truthfulness of a promise. If in the core, the promise is believable. If not, or if you're in doubt, be careful. Look for an outcome which definitely *is* in the core—in other words, find an outcome in which the promiser can't make a better deal with someone else.

We've applied this lie detector to a wide range of situations—in areas as diverse as love, politics, and donations to museums. However, none of the examples so far examined hinge on (or are supposed to hinge on) cash from the promiser's pocket to that of the promisee's.

This chapter also focuses on the core as a test of truth, but in a straight business deal.

* * *

Mo and Susan, married for seven years with two children, want to sell their house. Mo is a management consultant, associated with a medium-sized firm, Susan is a buyer for a large women's clothing store.

They have lived in this house all their married life, but can now afford to move to a bigger one in a more affluent neighborhood. The question is, how to go about selling?

Because used houses have been snatched off the market almost as fast as the sign is put on the lawn, Mo and Susan should be in a good position to sell. But they're not. Six months after buying this house, the state announced the construction of a new freeway. Their house was just outside the limits of property acquired by eminent domain. In other words, their place is easy to get to, but very noisy at rush hours.

What are their options?

First, they can try to sell the house themselves, putting a sign on the lawn, an ad in the paper, and a lot of time in front of the TV, sitting around waiting for a customer.

Second, they can list the house with a realtor, who presumably would do all of the work. A friend, Len, suggests a third option:

"Let me try to sell it. I don't have a realtor's license, so the final sale will be a private one. You say you want sixty thousand? Good. We'll split everything above that figure fifty-fifty. If sixty is all I can get, you take it all. However, the main point for you isn't so much the money, but that I'd put some energy into selling the place."

"Have you ever sold anyone's house before?"

"Only my own. I'm interested in seeing how well I do on someone else's."

Len advises them to look at the problem from the viewpoint of the potential buyer. A real estate agent will add five to ten thousand to the asking price just for bargaining purposes, an amount that will probably be knocked off at the end. In addition, the agent will add another 6 percent to cover his or her commission. So, if Mo and Susan hope to clear $60,000, the asking price would be about $75,000.

"With this listed price, you've already discouraged quite a few potential customers. They won't even come to look because the price seems too high."

Mo and Susan next contact a few real estate brokers, who look at the house. Each asks for an exclusive—signing a sixty-day contract with the broker. If sold during this time, the realtor gets his 6 percent regardless of who does the selling—whether Mo and Susan, Len, the broker, or anybody else.

"But, of course, I'll work very hard to sell this house. It's obviously in my interest to do so—the faster it's sold, the sooner I make my money. Besides, list with me, and you get a multiple listing; hundreds of other brokers in the area will also list the house. So thousands of potential buyers will immediately know about your property. It'll sell; you don't have to worry."

But they are worried. The realtor seems to have an incentive to sign them up, and then do nothing—he gains if the house is sold, period. If he thinks that Mo and Susan might peddle it, he may as well sit back and do nothing. Or, he may figure that another realtor will get into the act, in which case he still gets part of the commission.

When they raise this objection, he replies, "Suppose I bring over a customer who nibbles but doesn't bite. Six weeks later he comes back and buys. Or, he talks up the place to a friend, who then buys. I need protection. It's only fair."

Regardless of the route they go—and they have three from which to choose—they could sell the house, perhaps clearing their $60,000, perhaps settling for less, or they could fail to sell, see Figure 19.

They establish their priorities and walk-away strategy:

"Mo, if we don't sell it, what the hell difference does it make *how* we don't do it?"

"It makes a difference. If we list with a realtor, he can always use our place as the Brand X house to make some other dump look good. Or he might sit on the listing and do nothing."

"I suppose. But *we* might sell it."

"Yeah, but the realtor still gets his cut."

"Suppose Len tries but doesn't get anywhere?"

"We can always go to a realtor, Susan, or try ourselves. So it's not a complete strike-out."

"We might lose a good friend."

"Which would be too bad, but is always a danger if you do business with a friend. The point is he seems eager, and has no possible incentive to double-cross us. Why not let him take a stab at it?"

"I suppose we'll know if he really tries by the number of people he brings over."

Figure 19

No sale

Sale, bad price

Sale, Bonanza!

No sale

Sale, bad price

Sale, Bonanza!

No sale

Sale, bad price

Sale, Bonanza!

"Sure."

"But if we try ourselves and don't pull it off, we could still go to a realtor."

"So that's our walk-away strategy, which we could fall back on if Len didn't pull it off. My feeling is that we can at least count on his promise to try."

"Would we settle for less than sixty thousand?"

"Depends on how hungry we get."

"Well, selling it ourselves and getting a bad price is better than getting the same amount through a realtor."

"Of course, no commission. On the other hand, if Len gets us a bad price, he doesn't get a commission. That's his gamble."

"Mo, I'd rather someone else did the work, even if we get a good price. I don't want to be bothered."

Although Mo and Susan consider using a realtor, they decide instead to let Len try selling the place. A realtor may have a hidden motivation to make a better deal with someone else. They see two such possibilities: He can sit on his hands and hope that the multiple listing will eventually get another realtor to do the work. Alternatively, he could use their house as the example of what not to buy before showing a potential customer the real goods. Of course, the realtors they've talked to may well be straight shooters, but Mo and Susan have reasonable doubt. With Len there is no doubt about a hidden motivation. Since Susan doesn't want to be bothered with selling the house, a deal with Len is the most likely one to be in the core.

However, a realtor could bring along a high-rolling customer, who offers a lot of money, more than that offered by the humble bidders Len finds. But Mo and Susan can only count on this if the realtor actually brings the person over, not promises that he has such a buyer on tap. The realtor's credibility is based on what he delivers, not on what he promises.

* * *

This chapter has shown how to evaluate people's promises to see if they have a hidden motivation to make a better deal with someone else.

THE DEALS

First, list your options: the different possible people with whom you could make a deal.

FAILURE AND SUCCESS

Second, consider the possibilities that the deal will be a total failure. Also consider any degree of success you care to look at. In this example, Mo and Susan considered two levels of success—selling their house at a bad price and selling it at a good one.

DOUBLE-CROSS

Third, consider the possible reasons for a total failure with each person. If any include the possibility of a double-cross, you have good reason to doubt the person's promises. However, the person may be totally honest with you. You have merely raised the possibility that he or she is not being up front.

OTHER ANGLES

Fourth, quickly evaluate the remaining outcomes to see if there are overriding considerations which would counterweigh the doubts you have raised. In the example in this chapter, there were no such considerations.

THE CORE

Fifth, make the best deal which is least likely to run the risk of a double-cross.

You'll probably be in the core.

13

Who's on Your Side?

A psychotherapist: "I have a patient at the clinic who's collecting disability social security benefits. He's diagnosed as too emotionally disturbed to get a job. But I suspect he's faking it, and I think he realizes I'm on to him because our sessions are getting very, very tense."

One week later: "He did an amazing thing. He came into the office and asked, in fact demanded, that I bring in another therapist—anyone I wanted—to make a separate judgment on his condition. He's asking me to arrange for the staff to gang up on him."

The following week: "The other therapist and I thrashed out the whole thing; we decided that this man definitely was disturbed and needed to continue treatment. My colleague pointed out various ways I could better handle the situation."

Whether or not the patient was disturbed, he was

certainly no fool. Consciously or not, he understood the basic idea of coalitions, of two against one, and used this knowledge in reverse to lowball both therapists. Instead of trying to find a therapist who would side with him against the original one, he acknowledged the inevitable—therapists are members of a profession, and he's not one of them. So he apparently figured he was better off taking his chances by *relying* on professional solidarity.

Psychotherapists almost always tell you that they're humanists. A detective might look for fraud, but a humanist might look for the disturbance underlying a flimsy and transparent attempt to look like a fraud.

* * *

The principle of coalitions is very simple. Two or more people, each with separate strategies, deliberately coordinate them—for mutual gain, or to avoid maximum loss, or for humanitarian reasons. Sometimes the best combination looks bad for one member, but by making appropriate under-the-table adjustments—side payments—the problem can be smoothed out.

Often the cooperation is strictly against the rules, but happens anyway, as in criminal conspiracies, and in more everyday activities, such as auctions. Potential bidders may secretly agree to hold back with the aim of buying at the lowest price. One way to agree to do this was detailed in Chapter 11.

The reverse happens too. A common practice at modern art auctions is for the artist or his dealer to maintain the appearance of a stable floor under the value of the work. This is done with the help of a few friends scattered around the room. Their job is to make sure, in the absence of legitimate bidders, that the recorded bids will go almost, but not quite, up to the reserve price. As the final bids are usually printed in catalogues made up by the auction house, potential buyers can learn not only that the art work is still for sale, but that the bids are getting close to what the artist wants.

At one auction, two friends of the artist put up some spirited bidding against someone at the back of the room, who eventually offered slightly more than the reserve price and so actually bought the painting. Afterwards the artist and his friends went up to the auctioneer to ask who the buyer was.

"The back wall," he replied. So, the bidding was done by two confederates and one wall, which bought the painting.

Perhaps the major power of coalitions comes from the most seemingly trifling fact about them—the way the strategies of the members combine. Suppose each party in a two-member coalition has three options. How many do they have combined? Six? No, nine. Combined they have not the sum but the multiple of their separate options. To explain this: Suppose the first party uses strategy A. The second party can then use strategy A or B or C, whichever is best. If the first party uses strategy B, the second party can again use any of three strategies. And if the first party uses strategy C, the second party still has three strategies from which to choose—altogether nine possible strategies. Unpredictable outcomes turn into usable strategies, making the whole greater than the sum of its separate parts.

People who have collaborated on projects often comment about the "high energy level," a term that may seem nonsense or unscientific to those who haven't experienced it. Sometimes this phrase refers to a windfall of new ideas which the collaboration generates. Often, however, the ideas aren't really new, but are due instead to the geometrical rather than arithmetical growth of options. The variety was always there, but hadn't previously been apparent.

What about the person left out of a coalition? He's just where he started. If he begins with three options, he still has three. Now, however, he's up against a wall of possible strategies. He may have trouble enough guessing one person's strategies and values. To guess the strategies of two or more and what they can do together may simply be outside the limits of his ability. But the coalition, of course, has at least two people to zero in on him.

This confusion is sometimes turned by unscrupulous people to their advantage. For example, a buyer may arrange for a friendly business lunch with a seller. Arriving at the restaurant, the unsuspecting buyer may find not only the seller, but the seller's lawyer as well: "Jerry happened to drop by so I invited him along."

A similar gang-up happened to a husband a few weeks away from the divorce court. His wife asked him over for dinner in a last attempt at a reconciliation. "Just the two of us," she had told him over the phone. But when he arrived, who else happened to have called "unexpectedly?" Her psychiatrist, of course.

Indeed, a very successful operator in the museum world has advised: "Never go alone to any business meeting. Bring along the most impressive looking person you can, especially if he can keep his mouth shut."

The police use an interrogation gimmick based on coalitions—the Mutt and Jeff act. One cop is brutal and the other kind, the latter pretending to side with the suspect against the nasty cop. What is a poor crook to think?

All such gang-ups put the odd man out in the weak position of trying to figure out what he's up against, just as he's up against it.

What are office politics, but coalitions? They may be a drag, but one ignores them at one's own peril. For example, a Hollywood film was recently made about a famous, dead folk musician. Naturally, the star couldn't plunk a guitar without catching his fingernails on the strings. So Al, a well known musical figure and the world's authority on this folk hero, was hired to teach the star to fake it convincingly. Al would play the actual music off camera.

He was also the musical consultant. The idea was to bring unquestioned musical integrity to the project, and then plug this fact to high heaven in the promotion. However, Al came into conflict with the producer-director, who wanted the music played in a peculiar key. Al bravely objected: "The reason he played in this key is that this is the only one which works. No other key would sound like his music." He then drove the point home by recording the same piece in the two different keys, and letting the producer-director decide which was best. Since he had artistic pretensions, he couldn't help but agree with Al.

However, in this corner of Hollywood, money, not

quality, was the main consideration. The producer-director dreamed of adding an instrumental background, which did require a different key. The goal: market the soundtrack as a popular record conforming to the prevailing style, as opposed to the original one of the famous folk musician. "He had a unique sound, that's why he died broke," the producer-director kept saying.

Eventually, the production needed one more top-drawer musician, a banjo player. Al naturally hired the best he could find, but one who wasn't a close friend. You guessed it. This unscrupulous banjo player saw his opportunity, and took it. He didn't care what key he played the famous music in, and he could play the guitar almost as well as Al, whose movie career abruptly ended.

Sometimes a coalition helps one to face up to a weakness. For example, through mutual acquaintances a publisher heard about a proposed book a friend and I had talked about. He asked for a sample chapter, which we produced. His appetite whetted, he requested another chapter. We dutifully provided it. He then asked for a complete outline of the book. Again, we gave it to him. However, by this time we were tired of being nibbled to death with small requests. Either the publisher put up or shut up. We wanted a contract, and asked for it.

His reply: "Sure, I want to offer you a contract, but I never negotiate directly with authors. It's absolutely essential that we start off on the right foot. You guys go out and get yourselves an agent."

"What a professional guy, willing to deliberately increase our power relative to his." So we thought. The facts turned out to be just the opposite.

He correctly guessed that we would ask for a bigger advance than he or any other publisher would be prepared to pay. But an agent whom we would see as on our side could tell us the score—we were not only amateurs, but unknown ones.

So having a coalition on one's side may not mean victory, but it may mean doing as well as possible. Without the coalition, one might overplay one's hand, and do worse.

* * *

This chapter has shown the source of any coalition's power—the geometrical rather than arithmetical combinations of the members' strategies.

SOURCE OF CONFUSION
This fact underlies the confusion in the mind of someone who suddenly realizes that he's up against more than one person.

ALWAYS POSSIBLE
The possibility of coalitions should never be ignored, even if you consider such scheming to be particularly low-life.

OWN WEAKNESS
Sometimes the main purpose in forming a coalition is to become aware of one's own weaknesses. Lawyers and accountants are often engaged just for this purpose.

14

What Ever Happened to Mother-in-Law Jokes?

Recently, New York City's private nursing homes were the target of a major governmental and newspaper investigation. The snooping exposed a story that was shocking even for New York. Helpless patients were being systematically neglected, and in some cases, deliberately brutalized.

One man turned out to be the "czar" of most of the offending homes, controlling them through a network of paper organizations. He had connived at bringing about massive and avoidable suffering, but he managed to crawl through a few legal loopholes, and could only be prosecuted for perjury and Medicare fraud. Then, when he plea-bargained to a reduced version of even these charges, a lenient judge let him off with only a slap on the wrist—four months in jail.

Newspaper and television editorials gave voice to

what they claimed was the public's frustration. One network affiliate called the sentence a fraud, showing that white-collar crime pays. Undoubtedly, many people were outraged at the way he slipped off the hook, but was his punishment the main issue? Left unanswered in the whole shebang were some critical questions: If people really care about the fate of the aged, why are so many who could easily be cared for by their families sent to nursing homes in the first place? Why had the relatives of the victims ignored the issue for so long? Finally, if the relatives were personally insensitive, why hadn't social pressure forced them to do something?

This chapter shows how game theory handles standards of behavior. We'll see how to anticipate what might happen when they change.

* * *

Grandma, a widow, lives with her son (her only child), his wife, and their daughter, who is away at college. They share to a greater or lesser extent, their problems, hopes, dreams, and feelings, caring for each other with mutual emotional and financial support. Even Grandma contributes a social security check. It doesn't quite cover her expenses, but no one complains—her family loves her.

This is the ideal and is sometimes realized. However, it is being undermined by impersonal forces. Sociologists tell us that the modern family, no longer running the family farm or the family store, is now a unit of consumption. A glance at the ads on TV would tell us the same story. Rarely if ever, are the ads for items which allow the family to make money together. On the contrary, ads geared to families are for products they use up together—family toothpaste, family deodorants, family vacations, family station wagons, family swimming pools, family TV dinners, family this and family that. All these things cost money, so the fewer they go to, the more each gets. This is simply the nature of scarcity.

If other key components of family life are also in short supply, somebody in the family may be in big trouble.

Perhaps the son and daughter-in-law feel they don't have enough time and patience for Grandma. Sending her to a nursing home may be more expensive than keeping her at home, but her children may feel it's worth it. They may want to travel, and worrying about Grandma would cramp their style.

In any case, this isn't the only way the family might fragment. The daughter-in-law may protest Grandma's presence—out of jealousy, a money squeeze, or a personality conflict. The son may reply: "I have to take care of my mother; I simply can't send her to a nursing home."

"It's her or me."

"Where shall I forward your mail?"

Again, the issue is scarcity, not enough money, time, or love to go around.

The break-up could go the other way. The son may want his mother out of the way, but the wife, whose own parents are dead, may feel comforted by the presence of a mother figure. This is precisely what annoys her husband: "Why don't you grow up!" Feeling deprived of his wife's full attention, he leaves, and

Grandma and daughter-in-law are together.

Finally, all three might fight and end up going their separate ways. How could this happen? One possibility: The son asks his mother to leave, which she does. But, the daughter-in-law is so outraged over his callousness, that she leaves him. However, she's not sufficiently sympathetic to her mother-in-law to want to live with her. Or, if she is, Grandma may not go for it: "You're young, and should be around young people. I'll be okay in the home, around people my own age."

All of these situations happen every day. They are also the theoretical possibilities in a three-player coalition game. The coalition of all three is called the "grand coalition." Then, there are the three possible two-player coalitions. They exclude one player. Finally, there is the possibility of total fragmentation, with no coalition holding.

Only recently have all of these coalitions really been up for grabs. Earlier in this century the standard of behavior required children to take care of their aged parents, not as an act of charity, but simply as *the* way to handle the matter. This was the highpoint of the mother-in-law joke. For example, the open-air seat at the back of an enclosed car was called, among other things, a "mother-in-law seat." In spite of the jokes, only the crassest people would have sent Grandma packing. This would run the risk of losing the respect of friends, neighbors, and business associates: "If he'd throw out his own mother, what would he do to a mere boss?"

In game theory the old standard of behavior is called discriminatory. Grandma may have been given special privileges, but she lacked the right to talk herself out of the family. If she started to sound like a martyr and say that she wanted to go to the Old Folks' Home, she was politely but firmly told to shut up and drink her tea. Just look at 1930s movies on TV; the social rules are all there. Compare them with those implicit in Neil Simon's *The Sunshine Boys,* where two old vaudevillians both decide to finish their time at the Actors' Home. Their perfectly respectable families don't lift a finger to stop them.

Now the presence of Grandma in the family group is debatable. She's liberated, free. She may say she'll leave, or she may argue to stay. But she gets this "freedom" when she's in the weakest position to do either.

If haggling over scarce items is all that remains, we should look at the ultimate example. Its underlying logic may be more transparent but no less characteristic than that of milder situations.

The archetype appears in a grade B movie about an outlaw family who are so nasty they would "steal the gold in each others' teeth if they could agree on the split." As it turns out, Grandma has, after years of dental work, about $300 worth of gold in her teeth. Her son, having lived a shorter time, has less, only $200 worth of gold inlays. His wife had an easier childhood—a $100 gold cap. Is Grandma in big trouble? Not if she's a fast talker. If Sonny and his wife ask Grandma to say "Ah," she can quickly ask her daughter-in-law: "How much has he offered?"

"One hundred."

"Join with me against Sonny, and I'll make it one-fifty."

Obviously, Sonny has to increase his offer, or find a dentist fast. If they keep arguing over the split, with each making offers and counteroffers, stable bids probably would emerge. Grandma would be happy to settle for the gold already in her mouth; her coalition partner could have the whole take. If Sonny and his wife used the pliers on Grandma, the final split could be expected to be one hundred for Sonny and two hundred for his wife. The stable bids are those where each desperado in a coalition gets the same amount regardless of who the partner is. So no partner is preferred over any other, and we can't tell which coalition will form. But they never stop arguing, so they never get down to business.

For the desperado family the possibility of different coalitions shapes the demands. Although this may well be true of less gold-digging families too, it may not stalemate them. On the contrary, the anticipation of an endless argument may force the issue, and the argument may never happen.

Suppose the husband and wife are in the early stages of talking about what to do with Grandma. The husband wants to pack her off, but his wife seems to be leaning toward Grandma's side. He conjures up the fear that in some future fight his wife will side with Grandma against himself. Wasting no time, he runs out to pay a year's rent in advance at the local nursing home, presenting his wife with a *fait accompli.*

To sweeten the deal for her, the two of them go to Europe: "We couldn't have gone if we had to worry about Mother."

Grandma may have been free to bargain, but she was

never given the chance. The husband feared that his wife and Grandma would gang up on him. Although only a fantasy, this was good enough for him.

The entire game of social life is played in people's imaginations; only a few fragments become visible. Reality is a detail. Thus, in game theory, the cause of an event need never have happened. Carly Simon sings: "Anticipation, anticipation . . . is making me wait." But von Neumann and Morgenstern's song would go: "Anticipation is making me act."

What ever happened to mother-in-law jokes? They were replaced years ago with "What have you done for me lately?" jokes.

* * *

This chapter has shown how to anticipate the consequences of a change in the standard of behavior.

SCARCITY

First, determine if the situation revolves around scarcity.

THREE'S A CROWD

Second, check to see if there are at least three people in the situation.

BACKSTABBING

Third, if the first two requirements are met, all the ingredients are present for coalitions, arguments, and backstabbing.

THE JOKE'S ON YOU?

Fourth, see if there is a social convention holding back

the rawest consequences of unrestricted bargaining. If so, be grateful. If not, watch out, the joke may be on you.

ANTICIPATION

Fifth, remember that the future is not determined by what happened in the past, but on people's anticipations of what might happen in the future.

15

Left Out

The person left out of a coalition can do one of two things: abandon all hope or keep plugging away. The latter means working out the best walk-away strategy, which can be used either as a possible independent course of action, or as the basis for bargaining.

If you decide to keep up the fight, this chapter will help you to figure out a walk-away strategy.

Emma, seventy-nine, has been widowed for the past ten years. For the first nine she lived off her social security widow's pension, supplemented by savings. But, the cookie jar is empty now, and has been for the past year. This has forced her to move in with her son, Frank, a businessman, and his wife Alice, who have one daughter, now away at college. Emma has been given a small room with a private bath at the back of the house.

Her health is relatively good, and she is clearly not senile. But she does dwell on the past. Her reminiscing drives Frank crazy, particularly the endlessly repeated stories of his childhood. Seeing the way Frank is upset, Alice is beginning to get fed up with the stories too. They have asked Emma to stop telling them, but she persists.

For the past few months the situation has been very tense. Dinners, which had at first been jovial, and then at least pleasant, have now become heavy with silence. Emma suspects that her son and daughter-in-law have been secretly talking about packing her off to a nursing home. She has walked in on conversations that seem to be quickly changed when she enters the room.

"You'd never believe how often they talk about traffic jams," Emma has told her friend, Fay.

"I suppose this is Alice's doing?"

"That's the awful part. I don't think so. Maybe Frankie's afraid he'll lose her if he doesn't do something. But I suspect he's the one."

"Why?"

"It's always been that way. In the twenty years that I've known Alice, I've never once—not once—heard her make any suggestion, to do anything at all! They go to a movie, Frankie suggests it. They go out to eat, Frankie's idea."

"And Alice always agrees?"

"Sometimes not, I'll give her credit for that."

"What then?"

"Sometimes she says 'No, I won't do it.' More often she sulks, then gets bitchy, and is quite honestly, a pain in the behind. She goes along, but makes everyone miserable. And that's the way it's been lately."

"And why can't it be the reverse? Frankie talks about how nice it is to have you there, and Alice sulks?"

"It kills me to say it, but I think I've raised a selfish viper. He should be more patient, I'll be gone soon enough."

"Bite your tongue for talking that way."

"Would I have been any different? My own folks were lucky. They died young. Maybe I'd have been like Frankie."

"Emma, times were different."

"Times were, but people weren't."

"Your children love you, Emma."

"From a distance."

"Take Frankie aside and try to straighten it out. He's your son, talk to him."

"I should beg my own son to stay?"

"It's not begging. Okay, talk to Alice then."

"They're man and wife. I should come between them?"

"Then talk to them together. Find out what's going on. You've got to get hold of yourself, and deal with this thing logically."

Of course, Emma has never even heard of game theory, let alone a walk-away strategy, and she wouldn't know where to begin if asked to enumerate Frank and Alice's combined options. However, she's listed them without knowing it. Emma understands their general method of operation—Frank initiates and Alice agrees, sulks, or vetoes. Since Frank either wants Emma to stay or leave, and Alice will respond in one of three ways, there are six patterns of coalition behavior. Usually, the style of behavior expresses people's basic power relationships, which is just what you need to know in order to establish a walk-away value.

What are Emma's options? She has enumerated three of them—she can talk to Frank or Alice separately, or to both of them together. Over the next few days she reflects on her problem and realizes that she has two more options. One is to do nothing and take what comes. The other is to leave on her own without attempting to argue her case. Altogether, she has five strategies.

If Emma knew about game theory, she would merely have to make up a list such as that in Figure 20, pp. 178–79, examine each of her options, and decide which outcome was worst for each option. Then the option with the least bad outcome would be her walk-away strategy, and its worst outcome would be her walk-away value. Having established this, she could deal with her son and daughter-in-law knowing what she is willing to put up with. She'd have a base point.

Although Emma has never heard of game theory, she has a good idea of the actions open to her, and she has a feeling for the way her son and daughter-in-law behave. So she's virtually done a game-theory analysis without knowing it, and has worked out what amounts to a walk-away value, as she explains to Fay: "If I go on my own, and they don't try to stop me, I can tell them both to go to hell. It's not so bad."

"You said it."

"And if my son asks me to stay, but Alice says no, well I'm not the one breaking up their home, because I'm leaving. So, that's not so bad either. And if my son wants me out of his hair, but his wife Alice turns out to have more guts than I thought, good for her, but I'm leaving before the fight. If she's unhappy about it but keeps her mouth shut, well, it's disgusting, but *their* problem, because I'd be leaving. I wouldn't have to be around that ungrateful so and so, and his milk-toast wife. Altogether, going on my own leaves me with a shred of dignity. What else have I got, Fay?"

"Emma, aren't you going to stand up to them? You're just going to pack and go?"

"I'll tell them I'm leaving. If they raise a fuss, saying, 'Stay, we want you to stay,' and I see that they really mean it, I'll stay, but I don't want to stay under any other circumstances. Better to go away and die."

Why not evaluate all the outcomes as in previous examples? The final bargain will depend on exactly what's said, or not said, by everyone. So we don't know what to expect. But one's own walk-away value can and should be figured out *before* the talks begin.

Of course, Emma is only one of three bargainers in this situation. If Frank wanted to establish his walk-away strategy, he would consider his two options against the combined strategies of the opposing coalition of Alice and Emma. Since Alice has three options and Emma five, their coalition has fifteen strategies. As for Alice, she would look at her options against those of the combined opposition of Frank and Emma. Frank has two options, Emma five, so their coalition has ten options.

* * *

Figure 20
I'LL ACCEPT WHAT COMES.
Frank wants her to stay
Frank wants her to leave
I'M GETTING OUT OF HERE.
Frank wants her to stay
Frank wants her to leave
PLEASE, FRANKIE.
Frank wants her to stay
Frank wants her to leave
PLEASE, ALICE.
Frank wants her to stay
Frank wants her to leave
PLEASE, CHILDREN.
Frank wants her to stay
Frank wants her to leave

Alice { agrees / sulks / says no }

Alice { agrees / sulks / says no }

Alice { agrees / sulks / says no }

Alice { agrees / sulks / says no }

Alice { agrees / sulks / says no }

Alice { agrees / sulks / says no }

Alice { agrees / sulks / says no }

Alice { agrees / sulks / says no }

Alice { agrees / sulks / says no }

Alice { agrees / sulks / says no }

This chapter has shown how to figure out your walk-away value when faced with the possibility of being left out of a coalition.

OPTIONS

First, work out your options.

M.O.

Second, figure out the coalition's options. Unless you specifically know the details of what they have been scheming, you should look at their general method of operation because this is a tip-off to their power relationships. Recall that their combined number of strategies is the multiple of the number one person has, times the number the other has.

WORST

Third, for each of your options, decide which outcome is worst.

WALK-AWAY

Fourth, the strategy with the least bad outcome is your walk-away strategy, and the value to you of its worst outcome is your walk-away value.

16

Is Everybody Happy?

Most baseball team owners are also business tycoons, often self-made. But newspaper sports commentators commonly accuse them of being short on business sense, even if long on ego. The accusations may be justified. The typical pattern: after quietly making millions, the magnate buys a baseball team to get in the limelight. Phil Wrigley, owner of the Chicago Cubs is quoted as saying, "I could merge my chewing gum company with another business, and get an inch in the back pages of the *Wall Street Journal.* But if I suddenly sell a star player, I get on the front pages of papers across the country."

In 1976, Charlie Finley, owner of the Oakland Athletics, sold some of his best players in a banner-headlined three-and-a-half-million-dollar deal. Finley had to act fast. The players' contracts would soon expire, and once that happened they would be free agents, able to negotiate their own deals with any interested team—and without Charlie getting a cut. "Bad for baseball! But a good deal for himself!" screamed some sports commentators. Here was Charlie finally making a sharp baseball business deal, but he got it in the neck anyway. Should we pity the poor team owners? Better to pity the poor sportswriters. They struggle along on meager salaries and try to tell millionaire owners about the value of money, often suggesting that ego gratification is more reprehensible than is greed.

From our stance, the owners' priorities are their business. If they prefer headlines to dollars, fine.

As we have seen, the first requirement in making deals is to establish one's own priorities, and never to let anyone else say what they should be. You *feel* what you want; nobody else can do this for you. Sometimes the easiest approach is to start by looking at what you don't want, but are afraid you might get for each of your options. Reject choices which lead to the worst outcomes. Your walk-away strategy is the option with the least bad outcome, which becomes your walk-away value—the worst you need to get, if you play it smart. Use this strategy if you can't come to terms with anyone. You can also use it as a bench mark. Any deal should be better than your walk-away value. If not better, why bother with it? Even a quick use of this procedure should help you to deal with the situation more clearly.

If your aim is a dependable, long-term arrangement, the most frustrating person to deal with is one who doesn't know what he or she wants. So, paradoxically, the best thing you can do for yourself is to strengthen the other party's understanding of his or her own position. An example of this is in Chapter 13, "Who's on Your Side?"—a publisher encourages an author to get a literary agent in order to know the score and not make unreasonable demands.

If you take the opposite view—"keep 'em in the dark"—you may get the best of the situation (and the other party may even *deserve* to be taken, rather than rewarded for ignorance) but you may also make a long-term enemy. No one likes to be played for a sucker—friendships, families, and partnerships are hardly strengthened by such treatment. However, if you keep moving, you may not care. An example is the vendor who sets up late at night at subway stations and sells Mickey Mouse balloons. Hard-working people, finishing the night shift at menial jobs, buy them for their kids, feeling guilty that they haven't been able to spend the evening at home. Later, they discover that the balloons don't hold air. Naturally the vendor is not about to do business at the same station twice. If your business is flim-flam, you'd better keep moving.

What if you don't want to be a schemer, but the other party simply doesn't face up to the problem, and either can't or won't establish his or her priorities? Figure them out yourself; talking it over with a friend often helps. Try to put yourself in the other person's shoes, and see how he or she looks at the matter. Do this correctly and you know a tremendously important fact: If you don't want to corner the other guy, you know where the corner is. Always look at what the other guy wants and doesn't want.

If your goal is to leave everybody happy, you probably don't, and certainly shouldn't, try to squeeze the last drop of advantage from the situation. Instead, find an outcome which is better than the walk-away values for everyone in the deal. Occasionally, this is no problem, as in "Finessing It," Chapter 4, where the real problem was that one party, the dean, wanted the other, Fred, to take the lead. Then both could fulfill their wishes. The basic issue was that of signaling intentions. If your walk-away value is very high, you may as well start talking—you can only gain if you do.

What do you do if one partner has last-minute doubts about an arrangement, so both partners cannot get what they most want? The answer, again, is for the one with the clear course, and the high walk-away value to take the lead. The "negotiation" is really just a matter of making one's intentions clear. The one with cold feet may not be totally happy, but if he thinks about it and then goes along, he's certainly happier than he would be if he went his own way.

Social life is all about compromising. Often only one compromise is possible—take it or leave it. Game theory can be very helpful in clarifying and facing up to the situation, as in Chapter 5, "Giving Ms. van der Rohe the Business." Of course, if more than one compromise is possible, then there are other questions: "Why should *I* make the concession, why doesn't he? Or, why don't we both give in a little?" An example is in Chapter 8, "Let's Be Civilized," which deals with a divorce settlement. Forget about a "best" compromise, because who's to say what's best? Instead, the objective should be to focus on the limitations of each party, so that everybody is aware of all the potential damage. Then proceed both with caution and generosity.

A statesman is a disinterested promoter of the public good, as opposed to being a selfish promoter of his own interests. There is no comparable term for interpersonal agreements, but there should be. Something akin to statesmanship in personal life comes out of the game theory we've looked at, especially in Chapter 7, "Don't Take, Have Him Give." The whole point of a stable agreement is that everyone should be satisfied to the maximum possible extent. Sometimes this can be done at no personal expense but the party who could be generous takes a petty view of things: "Why should *he* get so much?"

The game-theory solutions we have advocated are always graciously generous. The basic point is to give the other guy the most you can consistent with the best you can do for yourself. Why advocate gracious generosity? Because generosity is moral? A good reason, but not the game-theory one, which is that generosity is *rational;* it leads to stability. Every statesman knows this, even if some politicians don't. Occasionally, even a statesman misses the point. Prime Minister Winston Churchill at first refused to allow the televising of Queen Elizabeth's coronation. Finally, he relented, but with one hitch: "I will allow this, provided no one at home gets a better view than I do myself from my stall in the Abbey."

The aim of office politics is the same as that of all coalitions—to exclude someone. Whatever is in short supply is then divided among fewer people.

The reason people are so often taken off guard by

coalitions is the enormous number of options open to those who can coordinate their strategies. Options don't merely increase, they mushroom.

When coalitions form, game theory's best advice is not directed to the winners, but to the losers, who can use the theory both to work out a best walk-away strategy, and to find the weakness in the opposing coalition.

For years, an old TV standby has been a program called "To Tell the Truth." A panel of celebrities tries to figure out which of three guests is the one all three claim to be. Most of us, at one time or another, have worried whether people were telling us the truth about one thing or another. It's only a serious problem when it concerns something we want. Several chapters have focused on a game-theory test for truth. The basic idea is to try to figure out if the person making the offer could make a better deal with someone else. If so, watch out.

Further Reading

NOTE: All of the examples in this book are based on true incidents, although, of course, the names of the persons have been changed. In many cases, the dialogue has been reconstructed from the memory of participants.

Robert Bell and John Coplans. *Decisions, Decisions: Game Theory and You.* New York: W. W. Norton, 1976. This book shows how to apply game theory to everyday problems in which people do not talk to each other, except maybe to try to pull a fast one.

Allan J. Cox. *Confessions of a Corporate Headhunter.* New York: Trident Press, 1973. If you want to find out more about headhunters than is described in Chapter 10, look here.

Morton D. Davis. *Game Theory.* New York: Basic Books, 1973. This is an excellent general discussion of the topic, but is not designed to help the reader in actual application.

Newton Frohlich. *Making the Best of It: A Common Sense Guide to Negotiating a Divorce.* New York: Harper and Row, 1971.

Paul B. Horton and Chester L. Hunt. *Sociology,* 3rd edition. New York: McGraw-Hill, 1972. The reference in Chapter 14 to the modern family as a unit of economic consumption can be found in just about any standard sociology textbook, for example, this one.

Chester B. Karrass. *Give and Take.* New York: T. Y. Crowell, 1974. An interesting list of bargaining gimmicks.

R. Duncan Luce and Howard Raiffa. *Games and Decisions.* New York: John Wiley, 1957. This is the standard reference work on the subject, and although highly technical, is generally not mathematical.

John McDonald. *The game of Business.* Garden City, New York: Doubleday, 1975. This is an interesting nonmathematical book, especially the sections on the core.

Oskar Morgenstern. "The Collaboration Between Oskar Morgenstern and John von Neumann on the Theory of Games." *Journal of Economic Literature,* September 1976. Although nothing in *Making Deals* is based on this very interesting article, it is a fascinating description of the collaboration of two brilliant men.

Gerard I. Nierenberg. *The Art of Negotiating.* New York: Cornerstone Library (paperback), 1968. The bargaining term "lowballing," used in Chapter 13, comes from this book. This strategem involves the appearance of both withdrawing from active opposition and of doing the reverse of what most people would do in a similar situation. Of course the term lowball comes from a variation of poker in which the lowest hand wins.

Anatol Rapoport. *N-Person Game Theory.* Ann Arbor: University of Michigan Press, 1970. This is a first-rate mathematical presentation of the subject. See this book for a full exposition of the differences between the theories of Nash and Shapley.

John Scanzoni. *Sexual Bargaining.* Englewood Cliffs, N.J.: Prentice-Hall, 1972. The divorce statistics referred to in Chapter 8 are from this excellent book.

Thomas Schelling. *The Strategy of Conflict.* New York: Oxford University Press, Galaxy Books (paperback), 1963. The discussion on threats in Chapter 9 is based on this book.

John von Neumann and Oskar Morgenstern, *The Theory of Games and Economic Behavior.* New York: John Wiley, Science Editions (paperback), 1964. The comments I wrote in *Decisions, Decisions* (New York: W. W. Norton, 1976) still apply: This is the classic work in the field. Although highly mathematical, the book was written with the aim of appealing to social scientists with little or no mathematical background. One year of college algebra (or even one semester) would be sufficient. The effort is well worth it, even now (more than), thirty years after publication. The authors lead the reader by the hand through the intricacies of the mathematics, and devote much of their writing to the motivation underlying the theory. The first forty-five pages should be read by anyone interested in game theory, psychology, or social science, regardless of mathematical background.

Edward Weisband and Thomas M. Franck. *Resignation in Protest.* New York: Grossman, 1975. Chapter 2 describes how many of our illustrious politicians exit from their jobs with shoes in hand, and by the back door. If you want to read on, this is one place to do so.